ARCHITECTS AND BUILDING CRAFTSMEN WITH WORK IN WILTSHIRE

An index compiled by Donald Watts, Barbara Rogers and others

Edited by Pamela M. Slocombe

Wiltshire Buildings Record
1996

First published in 1996 by
    Wiltshire Buildings Record
    Libraries, Museums and Arts Headquarters
    Bythesea Road
    Trowbridge
    Wiltshire BA14 8BS

Wiltshire Buildings Record

ISBN 0 9527933 0 X

Printed in Britain by
    Redwood Books
    Kennet Way
    Trowbridge
    Wiltshire BA14 8RN

Kennet District Council and the Wiltshire Branch of the RIBA have kindly assisted with part of the costs of publication.

This book is dedicated to the Wiltshire Branch of the Royal Institute of British Architects who have regularly given the Wiltshire Buildings Record encouragement and financial support since the Record was founded in 1979.

******

The Wiltshire Buildings Record is a voluntary society and educational charity, with members in Wiltshire and beyond. The archive of the Record, gathered together from fieldwork and from a variety of sources, covers many thousands of Wiltshire buildings past and present. The collection is currently open to the public on Tuesdays 9am-1pm and 2pm-5pm at Libraries, Museums and Arts Headquarters, Bythesea Road, Trowbridge, Wiltshire BA14 8BS. Telephone 01225 713740.

The Record has previously published three books (the first two in conjunction with Devizes Books Press and the third with Alan Sutton Publishing);

Wiltshire Farmhouses and Cottages 1500-1850
Wiltshire Farm Buildings 1500-1900
Medieval Houses of Wiltshire

Prices are £4.50, £5 and £6 respectively and the books are available direct from the Buildings Record or from Alan Sutton Publishing, Phoenix Mill, Stroud, Gloucs. GL5 2BU

CONTENTS

Page 5   Introduction

Page 7   Notes on the index, list of sources

Page 10   Index of architects and building craftsmen

Page 111   Photographs of architects from Dorling, 'Wilts and Dorset at the opening of the 20th century: contemporary biographies' 1906

Page 115   Index of parishes

Page 129   Advertisements from trade directories

Front cover: Model farm buildings and workshops at Stalls Farm, Horningsham on the Longleat estate from designs by W. Wilkinson of Oxford. Illustrated London News. Supplement 10.12.1859.

INTRODUCTION

This project has had a long period of gestation. It was initiated in 1984, the core work being carried out by WBR member, Donald Watts, a retired trained librarian, on index cards. His main sources were Pevsner's Buildings of Wiltshire, the volumes of the Victoria County History of Wiltshire and the Diocesan records at the Wiltshire Record Office. The WRO had already indexed the architects represented in the records themselves. Don moved on to look at many other sources and eventually reached a point where he felt he should call a halt as access to some buildings with archives, and travel, were difficult for him. The card index was passed over to the WBR office and from the beginning proved a valuable asset.

Julian Orbach, a well-known architectural historian, who was leaving the county, then generously supplied us with a large number of additional references from his searches of old copies of 'The Builder' and other sources. Barbara Rogers, a WBR committee member and volunteer added them into the index. My daughter Elizabeth Slocombe began the work of typing the index on to a word processor, most of which was completed by Doris Roddham, another volunteer. Since then I have made gradual additions to the index as new information has come to light. Robert Gutchen's study of Wiltshire workhouses has been especially useful and my husband Ivor Slocombe searched the records of government grants to schools at the Wiltshire Record Office. These, incidentally, are accompanied by many fine plans. I have given the lists a final editing and added a second index of parishes.

The fascination of the index lies in the drawing together of work by one person in different parishes, the demonstration of family involvement often over a long period of time and the merging and development of categories of work; mason, quarry owner, builder, surveyor, land agent, architect, artist. Until fairly recent years, these roles were linked and interchangeable and a person might graduate from one to another in his lifetime. There is often a tendency too in families to move from one trade or profession to another within the building sphere. For example, a mason might have a brother who was a carpenter and they could carry out work together or a builder father might have an architect son. Andrew White in an article in 'The Local Historian' (May 1991) has pointed out that it may have been drawing skills which distinguished the architect from the builder in the Georgian period. Architects often did surveying work between building commissions and often an architect produced a specification and a front elevation, leaving the builder to design the rest. We see also in the index the varying styles produced by a single person and we see how Wiltshire can boast work by many nationally famous people. Lastly, I hope this index will do something to break down the artificial boundaries between the studies of 'polite' and 'vernacular' architecture.

The index is, of course, still in its infancy. So many interesting buildings in the county have unknown builders and architects and it should be especially easy to trace names connected with large houses, town halls, chapels, hospitals etc. Many names are buried in letters and other documents at the Record Office, in old newspaper items and sale particulars and in private collections of deeds and papers. From the late 19th century onwards there are planning applications at WRO. There are further local trade directories which could be searched. Much could also be discovered by questioning elderly members of building firms. Local historians could play a useful part in this. However, we hope this list will prove useful and arouse your interest to contribute new entries and biographical details towards a lengthier edition or supplement in the future. Corrections will also be very welcome.

Pamela M. Slocombe

NOTES ON THE INDEX

Dates
Under each entry examples of work are in date order where this is known. Sources sometimes disagree slightly about a date because of the time taken between drawing the plans and completion of the building. Where the date of the plans is known this has been preferred. Undated entries have not been in any way put in order by style etc. A decision was made to omit in general entries after 1960 (though the card index in the office continues to the present day) because of the difficulty in obtaining comprehensive lists of modern work and a desire not to be unfair to practising architects. Exceptions to this occur when a few buildings after 1960 come at the end of an architect or craftsman's work.

Parishes
Places are all listed under the civil parishes of Wiltshire in being in 1996. Where two parishes have the same name N (north) or S (south) is used to differentiate them.

Entries
Some names have been included without any buildings being mentioned when the person or firm was resident in the county, in case local work by them is discovered.

LIST OF SOURCES

A source is given for most entries but in a few cases a record card was made omitting the source.

ADK A.D.Kirby, article in Wyvernacular newsletter 1965
AS Ann Smith 'Sherborne Castle: from Tudor Lodge to Country House', article in The Local Historian, Nov 1995
B Edward Bradby, 'The Book of Devizes', Buckingham, Barracuda Books, 1985
BB B. Barrett, local historian, Rowde
BC Barry Cunliffe. Article in Bath History, vol. 1, 1986
BFC Basil F.L.Clarke, 'Church builders of the Nineteenth Century' 1969, David and Charles
BG Julian Orbach, Blue Guide 'Victorian Architecture in Britain', London, A & C Black, 1987
BM Robert (Bob) Machin, course tutor, vernacular architecture
C H.M Colvin, 'A biographical dictionary of English architects 1660-1840', London, John Murray, 1954
CC Chippenham Civic Society, 'A Chippenham Collection', Chippenham Civic Society, 1987
CH 'Chamber's Biographical Dictionary', Chambers, 1988
CS Christopher Stell, architectural historian
D Rev E.E Dorling, 'Wilts and Dorset at the opening of the 20th century: contemporary biographies', Brighton, 1906
DH Mrs Diana Holmes, local historian, Latton
DNB Concise dictionary of national biography. Part 1: from the beginnings to 1900. Part 2: 1901-1950
DOE Department of the Environment (now Dept. of National Heritage) listed buildings descriptions
DR Dauntsey Rectory papers, at house and WRO

EG	Elizabeth Gibb, WBR member, Devizes
ETD	'Early Trade Directories', Wiltshire Record Society
FH&P	John Fleming, Hugh Honour and Nikolaus Pevsner,'The Penguin dictionary of architecture', Harmondsworth, Penguin, 2nd ed., 1972
G	Roderick Gradidge,'Dream houses: the Edwardian ideal', London, Constable, 1980
GB	Geoffrey Butcher, builder
GL	Gee Langdon, 'The Year of the Map', Compton Russell, 1976
GS	Gavin Stamp, architectural historian
GW	Giles Worsley, 'Architectural Drawings of the Regency Period'
HA	Halliday notebook (WRO 2735)
HF	Harold Fassnidge,'Bradford-on-Avon. A Pictorial Record' Wilts Library and Museum Service, 1983
HH	'History of Holt United Reformed Church'
HHF	N.M.Morland, 'Houses and Housing in Wilton 1831-1993'
ifo	Information from owner
II	Isabel Ide,'The Celebrated Melksham Spa', 1988
JB	John Betjeman. Article in 'Studies in the History of Swindon', Swindon Borough Council 1950
JF	John Farrant, SWIAS member, article in Search No 7, Autumn 1967
JMR	J.M.Robinson, 'The Wyatts', 1979
JNH	J.N.Haresnape, local historian, Pewsey
JO	Julian Orbach, architectural historian, notes for visits by Victorian Society Avon Group to places in Wiltshire and personal notes compiled from copies of 'The Builder' and other sources
KAR	Kirsty A. Rodwell, structural archaeologist, Box
KP	Kenneth Powell, article in Daily Telegraph 15.8.1991
KR	Kenneth Rogers, former County Archivist, Wilts
L	Lewis, 'Church Rambler', 1876
LB	Lawrence Burton, 'A Choice Over Our Heads', London, Talisman Books 1978
LP	Long Papers, WRO
M	Rev. W.J.Meers, 'Kington Langley', 1937
MB	Marion Barter, architectural historian involved in the DOE resurvey of listed buildings in Wiltshire
MG	Marc Girouard, architectural historian
MJ	Michael Jones (of Swindon?)
ML	Michael Lansdown, local historian, Trowbridge
MM	Margaret Maxwell, architect, Pewsey
MMa	Michael Marshman, local studies librarian, WCC
MP	Methuen Papers, WRO
MS	Michael S. Smith, architect, Bradford-on-Avon
MW	Martin Watts, 'Wiltshire Windmills', Wiltshire Library and Museum Service 1980
NJ	Neil Jackson, 'Nineteenth Century Bath Architects and Architecture', Ashgrove Press, Bath 1991
NMR	National Monuments Record Centre, Kemble Drive, Swindon
NT	National Trust guidebook
P	Nikolaus Pevsner,'Buildings of Wiltshire', Harmondsworth, Penguin, 2nd ed., revised by Bridget Cherry, 1975
PE	P.W.Edwards, article in first Wyvernacular newsletter, Christmas 1964
PL	Ed. Michael Cowan, 'The Letters of John Peniston, Salisbury Architect, Catholic, and Yeomanry Officer 1823-1830', Wiltshire Record Society 1996

PMS	Pamela Slocombe, Organiser of Wiltshire Buildings Record 1979-1995
PP	June Badeni, 'Past People in Wiltshire and Gloucestershire', Badeni 1992
PW	Paul Woodfield, architectural historian involved in the DOE resurvey of listed buildings in Wiltshire
R	'A good job well done: the story of Rendell, a west country builder', Amersham, Stocker Hocknell, 1983
RA	Richard Atkinson, 'Alderbury - an ancient and peculiar parish'
RC	Ross Churn, Hatfield, Herts, 'The Vanished Dwellings of Dilton' unpublished entry in Wiltshire Life Society local history competition 1992/3
RCHM	Royal Commission on Historical Monuments of England, 'Ancient and historical monuments in the City of Salisbury', London, HMSO, vol 1, 1980. 'Wilton House and English Palladianism' HMSO, 1988. 'Nonconformist Chapels and Meeting Houses - S.W. England' and various reports on individual buildings.
RG	Professor Robert Gutchen, University of Rhode Island, USA
RGr	Rosemary Green, 'Bibliography of Wiltshire 1920-1960' WCC
RH	Richard Hatchwell, antiquarian bookseller and old print specialist
RIBA	Royal Institute of British Architects, RIBA Directory of practices, 1984
RKN	Ronald K. Newman, former nurse, Roundway Hospital
RM	Roger Mawby, local historian, Bradford-on-Avon
S	Alan T.C Slee,'The building history of Urchfont Manor', WAM vol 79 1985 pp192-200
SA	Susan Arnold, curator, Great Barn Museum, Avebury
SC	Sale catalogue
SHS	Sheila C. Povey, 'Swindon Ex-servicemen's Housing Society, Farleigh Crescent, 1954-1963' unpublished entry in Wiltshire Life Society local history competition 1992/3
SJ	Sheila Judge, local historian, Easterton
SP	Sale particulars
SPC	Rita Marshall, 'History of St Paul's Church, Chippenham' c1990, unpublished
TM	Tim Mowl, architectural historian, involved in DOE resurvey of listed buildings in Wiltshire
TS	Tom Smith, WBR member, Swindon
VC	Victory Chinnery, 'Oak Furniture', Antique Collectors' Club Ltd., 1979
VCH	'Victoria County History of Wiltshire', 12 vols, London, OUP, for the Institute of Historical Research
VH	Ed. A. Langley and J. Utting, 'Village on the Hill' 1990
VH2	Ed. J. and J. Utting, 'The Village on the Hill Vol. 2', Colerne History Group, 1995
WAM	Wiltshire Archaeological Magazine
WBR	Wiltshire Buildings Record archives, Trowbridge
WG	Wiltshire Gazette
WR	Warwick Rodwell, church archaeologist
WRO	Wiltshire Record Office (Index of architects mentioned in Diocesan records and numerous other sources)
WT	Wiltshire Times
WWA	'Who's Who in Architecture 1926'
YP	'Fifty Years of Progress in Bradford-on-Avon' 1887

INDEX OF ARCHITECTS AND BUILDING CRAFTSMEN

ABINGTON, L.J architect
1. DEVIZES. Market Cross (with B. Wyatt) 1814   P

ABRAHAM, Robert A architect 1774-1850
Most of his work was for the Catholic aristocracy.   C
1. MILDENHALL. Former School at E end of village (now a private house)
   1823 P   1824 VCH
2. ODSTOCK. Longford Castle, plans for extension (only partly carried out)
   P

ADAIR, John joiner
1. WILTON. Wilton House. Doorcases etc c1761   RCHM

ADAM, Robert architect and designer 1728-1792
Born at Kirkcaldy, son of William Adam, architect. Established practice in
London in 1758 and worked with brother James. Classical design, interior
decoration and furniture and fittings.   CH
1. CALNE WITHOUT. Bowood. Additions with James Adam in the 1760s, inc. S
   portico; mausoleum, 1765; dining room, centre portico, orangery,
   library, 1769-70 (after 1955-6 south range only remains of Adam's work)
   WAM 41, 1932   P
2. CALNE. Castle Street, Castle House - back range, 1770 (now old people's
   home after gutting in 1950s)   P
3. FONTHILL GIFFORD. Fonthill Abbey. Designs for the house 1763   P

ADYE, Charles Septimus architect 1841-c1911 of Westbury House, Bradford-on-Avon
Articled pupil of Manners & Gill, city architects of Bath. First County
Surveyor of Wiltshire County Council 1889-1906. Had previously worked part-
time for Quarter Sessions. KR By 1895 practice included Charles Septimus
and Herbert Archibald Adye. D C Woolley Adye, Town Hall Chambers, Bradford-
on-Avon (Kelly's 1875, 1880, 1889). Work for W.C.C. included new police
stations, new County Offices etc 1889 on. P Church restorations and new
building all over Wessex, and in Essex and Derbyshire. Photograph in D.
1. BRADFORD-ON-AVON. National School, urinals 1867   WRO
2. BRADFORD-ON-AVON. Frome Road, St Catherine's Almshouses 1868   HF   YP
3. BRADFORD-ON-AVON. Church Street, Glebe Cottage (now called Orpin's
   House). Plans for enlargement and use as National School school master's
   house 1869   WRO
4. BRADFORD-ON-AVON. Frome Road, Hall's Almshouses (of 1700), restored
   1870-80   JO
5. KEEVIL. St Leonard's Church, alterations 1874-6   WRO
6. MONKTON FARLEIGH. St Peter, restoration work, 1875?   WRO
7. MELKSHAM. Forest, St Andrew 1876   P JO L
8. KEEVIL. Talboys, restoration 1876-80
9. BRADFORD-ON-AVON. Temperance Tavern (Knee's Corner), 1878-9 (opened
   8.8.1879)   JO
10. BRADFORD-ON-AVON. Restoration of Saxon church of St Lawrence, after J.T.
    Irvine. Adye was responsible for the present W wall 1880s   P
11. ROUNDWAY. County Lunatic Asylum (later Roundway Hospital), chapel 1895
    (drawings, but not built), additions 1898 and 1903 (possibly new

laundry, conversion of isolation hospital to villa and erection of
adjoining 54-bed villa)   RCHM

ADYE, Henry Archibald architect
Practice with C.S.Adye at Town Hall Chambers, Bradford-on-Avon (Kelly's 1889).
1. BRADFORD-ON-AVON. Newtown, Church Schools 1896 (now demolished)   JO

AINSWORTH & PILCHER architects
Office in Regent Circus, Swindon (Kelly's 1907)

AITCHISON, George, Junior architect of London
1. PITTON AND FARLEY. Farley. National Schol 1868   WRO
2. EAST KNOYLE. National School 1872-3 (for Alfred Seymour of Knoyle
   House)   DOE   VCH

ALEXANDER, A & SON builders
Demolished The Priory, Market Street, Bradford-on-Avon, 1938.   WT

ALEXANDER, Daniel Asher (David in P) architect and engineer 1768-1846
1. ODSTOCK. Longford Castle. Alterations 1802-17 (after James Wyatt's
   designs). Only north tower rebuilt and two new sides started   P
2. DOWNTON. St Laurence, alterations to nave and aisles (for Lord Radnor)
   1812-15   P

ALEXANDER, George architect or builder
1. HIGHWORTH. Infants School 1866   WRO

ALFORD, John builder
1. SUTTON MANDEVILLE. Chickgrove. School and house 1874   WRO

ALLEN, James Mountford architect 1809-1883
Practised in London, later, as a church architect, in Crewkerne, Somerset
DNB
1. WEST KNOYLE. St Mary's Church 1878 (except perp. tower)   P

ALLOM, Thomas architect and artist 1804-1872
Illustrator of travel books.   DNB C
1. CALNE. Workhouse (design exhibited Royal Academy 1847)   JO

ANGELL, Thomas architect of Highworth
1. LATTON. Parsonage 1826   WRO
2. STRATTON ST MARGARET. Parsonage 1827   WRO
3. HIGHWORTH. Old workhouse alterations   RG

ANREP, Boris
1. LUDGERSHALL. Biddesden House. Changing pavilion for pool, decorative
   mosaics 1937 VCH

ANSTIE, John clothier of Devizes 1743-1830
1. DEVIZES. New Park Street, clothing factory 1785 (designed and specified
   by Anstie)   WBR   P

ARCHITECTS CO-PARTNERSHIP
A group of idealistic left-wing architects all born about 1915-7. Best
known work - factory at Bryn Mawr (1949), "very beautiful". GS
An office at 6 Wood Street, Swindon (head office at Potter's Bar) FH&P
1. SWINDON. Raychem factory   P

ARMSTRONG, John builder
1. SWINDON. College Street. Girls and Infants School 1873   WRO

ARNOLD, William mason-architect of Somerset
Worked in Wilts., Dorset, Somerset and further afield. Active late 16th-
early 17thC.   KAR
1. STOCKTON. Stockton House   KAR
2. ANSTY. The Hospice 1594-8? (probable attribution)   KAR

ASH, Henry builder of Devizes
Firm established 1830. Took over inn and brewery site at 13 Bridewell
Street after closure in 1866. Builder's yard behind house. Maslen q.v.
succeeded on site. In 1906 firm was based at Northgate Street and
Commercial Road (Mate's Illustrated Guide of Devizes). WBR
1. DEVIZES. 29 Market Place, offices for Anstie's (architect Randell) 1894
   WBR
2. DEVIZES. Bath Road, Hazel Rock and The Gables, semi-detached pair. By
   1906   WBR

ASHBEE, Charles Robert architect and designer 1863-1942
1. CALNE. St Mary's Chapel, reredos and organ case 1907   WRO

ASHLEY, Edward architect
1. DURNFORD. School and house 1872   WRO

ATKINSON, Thomas architect of Salisbury
1. UPTON LOVELL. Parsonage 1802   WRO
2. SALISBURY. General Infirmary, plans in 1819, may have carried out
   works.   RCHM

ATWOOD (See **JONES AND ATWOOD**)

ATWOOD, Thomas Warr architect
One of a noted Bath family of builders and developers. Killed 1775. Elder
brother lived at Turleigh Manor, Turleigh, Winsley 1775-1808   JO

AUST, David builder of Bath
1. LIMPLEY STOKE. Viaduct 1834 (G.P Manners, architect)   JO
2. DILTON MARSH. Dilton Court 1842 (G.P.Manners and J.Peniston architects)
   Owner's plans

AUSTIN, SHOUT & WITHERS architects of Bristol
1. DILTON MARSH. Parsonage 1852   WRO

AWDRY, Graham C architect 1858- of Westminster   WWA JO
1. CHIPPENHAM. Lowden Mission Hall 1885   WT
2. CHIPPENHAM. London Road, Cottage Hospital, opened 1899   JO
3. DEVIZES. Almshouses late 19th C   JO

4. MALMESBURY. Cemetery late 19th C   JO
5. LUDGERSHALL. St James. Restoration work 1900   WRO

AYLMER, Guy architect
1. ALTON. West Stowell. House. Remodelling and enlargement c1930   VCH

AYLWIN & MAY of Marlborough and Newbury
1. MARLBOROUGH. St Peter's National School. Site plans 1850   WRO

BACK, Edward H architect
Practice at 15 Long Street, Devizes (Kelly's 1880)

BAILEY, E.N and PARTNERS
1. SWINDON. Cheney Manor estate, Remploy building 1958   P

BAILEY, John freemason of Westwood
An heir and executor of will of Mary Silverthorn (nee Bailey?) of Hilperton 1728.

BAILEY (BAILY or BAYLY), Roger senior mason of Westwood
1. SEMINGTON. Whaddon House and estate. Work (stone and brick) 1673-1680
   (eg 113 days at 16d in 1680)   LP

BAILEY, Roger junior mason of Westwood
Son of Roger Bailey senior.
1. SEMINGTON. Whaddon House and estate. Work 1678-80 (eg 51 days at 14d in
   1680)   LP

BAILEY, Samuel mason of Westwood
Son of Roger Bailey senior.
1. SEMINGTON. Whaddon House and estate. Work 1680 (121 days at 10d)   LP

BAKER & HINTON
Orlando Baker. Office at 38 Regent Street, Swindon (Kelly's 1875,1880,1889)

BALDWIN, Thomas architect 1750-1820
Began as a speculative builder in Bath. City Surveyor in 1775, Deputy
Chamberlain and Surveyor 1785. Planned Bathwick New Town, later taken over
by John Pinch. NJ Bankrupt 1793, but continued to practise as an
architect.   C
1. DEVIZES. Town Hall 1806-8   P
2. WILCOT. Oare, Rainscombe House (ascription), soon after 1816   P

BALL family thatchers of Avebury   SA

BARFIELD, T.H architect or builder
1. INGLESHAM. School and house 1871   WRO

BARKER, E.H. Lingen architect of Hereford
1. CODFORD. St Mary. Restoration of 1878-9   P

BARNDEN, John architect of Warminster
1. WINGFIELD. St Mary. Restoration work 1861   WRO
2. EBBESBOURNE WAKE. Fifield Bavant. Parsonage 1867   WRO

BARNES, Edward plasterer
1. DAUNTSEY. Rectory 1829-33   DR

BARNES, Jacob plasterer
1. DAUNTSEY. Rectory 1829-33   DR

BARNES, John carpenter
1. DAUNTSEY. Rectory 1829-33   DR

BARNES, Joseph carpenter
1. DAUNTSEY. Rectory 1829-33   DR

BARNES, R architect of Dinton
1. TEFFONT. Teffont Evias. Vicarage altered or rebuilt 1805   JO

BARNES, William architect of Dinton
1. SUTTON MANDEVILLE. Parsonage 1832   WRO

BARRETT, James mason of Kington Manor, Kington St Michael   -1782
His brother Charles also involved in firm. PP Possibly owned quarry. PMS
1. KINGTON LANGLEY. Greathouse. Work on privy 1752   PP
2. KINGTON LANGLEY. Mausoleum of Coleman family at Chapel Field 1782 PP

BARRETT, Thomas architect of Swindon
1. WOOTTON BASSETT. British School 1859   WRO

BARRY, Sir Charles architect 1795-1860
Gothic, Grecian, neo-Renaissance in succession. Houses of Parliament
competition won 1835-6, begun 1839, opened 1852   FH&P   His most
successful work was his Italianate style.   GW
1. CALNE WITHOUT. Bowood. Tower 1840 and Golden Gates 1834-1857   P
2. CHERHILL. Obelisk 1843-5 to commemorate William Petty   JO

BATEMAN, William Herbert architect
Practice at Mill Street, Calne (Kelly's 1927)

BATH, Frederick architect 1847-? of Sandown House, Churchfields,
Salisbury
Practice at Market Square, Salisbury (Kelly's 1875); Crown Chambers, Bridge
Street in 1880, and practice still there at least to 1915. Photograph and
biography in D.
1. SALISBURY. New Canal. House of John Hall, new front 1880-1, (the cinema
   of 1931 incorporates as its entrance this house)   VCH
2. SALISBURY. Palace Theatre 1888 (now demolished)   P
3. SALISBURY. County Hall   D
4. SALISBURY. Fisherton Schools   D
5. SALISBURY. Milford Manor   D
6. SALISBURY. New Sarum House   D
7. SALISBURY. Bloom's premises   D
8. CALNE. Wilts and Dorset Bank   D
9. SALISBURY. Fisherton Anger, St Paul. Restoration work 1917 WRO

BATTEN, Isaac mason
Paved lane on N side of Trinity Church, Bradford-on-Avon 1779. GL Mortgagee for Church Street woollen mill in 1810. KR

BATTEN, James carpenter, joiner and builder of Bradford-on-Avon
At Bearfield in 1830. ETD. Owned and occupied various property in the town in 1841. Was possibly the James Batten, carpenter, joiner and builder at Woolley Street in 1822-3. ETD

BATTEN, Jeremiah builder of Bradford-on-Avon
Mason at White Hill in 1822-3 (ETD), will 5.2.1830, dead before 1862 (private deed). Owned probable quarry W. of Belcombe Court. PMS

BATTEN, Robert mason of Bradford-on-Avon
1793-8 ETD

BAVERSTOCK, James architect, carpenter and joiner of Marlborough
At St Margaret's in 1842. ETD
1. OGBOURNE ST ANDREW. All Saints 1872 (closed 1961)   VCH

BAVERSTOCK, William Edwin architect, builder and carpenter
Practice at High Street, Marlborough ETD 1842 and Kelly's 1875-1895.
1. MARLBOROUGH. St Mary's. School and house 1850   WRO
2. CHISELDON. Vicarage 1861   WRO
3. BROAD HINTON. Vicarage. Partial rebuilding 1867 (sold 1978) VCH
4. WOOTTON RIVERS. School, addition 1870   WRO
5. OGBOURNE ST ANDREW. School and house 1871 (architect Pope) WRO
6. WANBOROUGH. School, additional schoolroom 1872   WRO

BEAN, W.J architect of London
1. BERWICK ST JOHN. Vicarage 1880   WRO   JO

BECKETT, Archibald architect
1. TISBURY. Conversion of workhouse to brewery 1868   RG

BECKHAM, Benjamin joiner of Salisbury
Brother of Humphrey who left him his workshop and tools in 1671. VC

BECKHAM, Humphrey joiner-carver of Salisbury 1588-1671
Born Salisbury of clothier parents. Apprenticed to Mr Rosgrave, painter and carver c1605. Made furniture as well as fittings and carved ornamental work on buildings. VC
1. SALISBURY. Brown Street, carved work in Mr Thomas Dennis's parlour c1600
   VC
2. SALISBURY. St Ann Street, Joiners' Hall. Elaborate carvings on facade popularly attributed to Beckham. Early 17thC   VC
3. SALISBURY. St Thomas's Church. Beckham's own carved oak tomb panel, probably made as chimney overmantel in 1620-40

BECKHAM, John joiner of Salisbury
Brother of Humphrey. Member of Joiners' Company 1615-22 at least. VC
1. SALISBURY. Council House. Wainscot and carved works 1634   VC

BECKHAM Nathaniel joiner of Salisbury
Relative of Humphrey. Alive 1671. VC

BECKHAM, Reginald joiner of Salisbury
Freeholder of city in 1607-8. WAM 19

BECKHAM, Reynold joiner
Same as Reginald?
1. SALISBURY. St Thomas's parish, Queen's Arms Inn, paid for frame 1573 VC

BECKHAM, William joiner of Salisbury
Brother of Humphrey. Mentioned in Joiners' Company accounts 1618-21. VC

BELCHER, John architect 1841-1913
Gothic at first, then Italian Renaissance style. Designs include V&A Museum, 1891, Whiteley's Stores, 1912, etc.   DNB
1. SOUTH MARSTON. St Mary Magdalen - nave, chancel, arch, etc 1886 P

BELL, Joseph glazier of Bristol
1. TROWBRIDGE. St James's Church. Glass when restored 1847-8 and in 1859 JO

BELLAMY, Thomas architect
1. CORSHAM. Corsham Court. N. range to replace Nash's work there 1845-9 P

BENSON, William 1682-1754
Held lease of Amesbury Abbey in 1709. RCHM Appointed High Sheriff of Wiltshire 1710. MP for Shaftesbury 1715 C Surveyor General for a short time after Wren's dismissal. P
1. NEWTON TONEY. Wilbury House. Benson bought the land and designed the house himself in 1710, from one of Inigo Jones's unpublished drawings. The house is an early example of neo-Palladian style.   RCHM   P C
2. STOURTON WITH GASPER. Stourhead. May have been involved there, brother-in-law of Hoare and the two men collaborated on a church chancel in Hants. in 1723   RCHM

BENTLEY, J.F architect
1. SUTTON BENGER. Draycot Cerne, Upper Draycot. Cemetery chapel, opened c1883 VCH

BERRETT, thatcher of Hilperton?
1. SEMINGTON. Whaddon House estate. Farmhouses and oxhouses etc. 1675-1677 LP

BERTODANO, H.S de architect
Practice at 5 Bloomsbury Square, London, WC1
1. HANKERTON. Parsonage, extensions 1903 and new Vicarage (later called Hankerton Priory) 1905-6   VCH   WRO

BERTRAM, BERTRAM and RICE architects of Oxford
1. SWINDON. Euclid Street. Civic Offices 1936-8 (Two quadrangles) VCH vol9

BESSANT, John master mason
1. DAUNTSEY. Rectory 1829-33   DR

BESWICK, A.E architect of Swindon
Son of R.J.Beswick. Firm of R.J.Beswick & Son. Ran it single handedly when war broke out in 1939. ADK
1. MILDENHALL. St John. Restoration work 1936   WRO

BESWICK, R.E.E architect of Swindon
Son of A.E. Beswick. Partner in R.J Beswick & Son of 10 Victoria Road, Swindon. Trained at Liverpool School of Architecture. Modernist. Became Lieutenant in Army during the 1939-45 war. ADK
1. CHISELDON. Holy Cross. Restoration work 1934, 1938   WRO
2. BROAD HINTON. St Peter ad Vincula. Restoration work 1936 WRO
3. SWINDON. Raleigh Avenue. St Andrew 1957-8   P
4. SWINDON. Farleigh Crescent. Self-build houses. Nos 16 evens to 22 completed 1958, 24 evens to 30 1959, 32 and 34 1959/60. SHS
5. SWINDON. Ashford Road. St Saviour 1961 (the wooden church of 1890 was encased in stone, with new vestries, by Beswick).   P

BESWICK, Robert James architect
Practice at Fleet Street, Swindon (Kelly's 1889-1895); at 9 Regent Street (Kelly's 1899); at 10 Victoria Road, Swindon (Kelly's 1907-1927 at least). Later became R.J.Beswick & Son. This firm joined with Edwards & Webster to form Wyvern Design Group in 1952. PE
1. SWINDON. Regent Street. Baptist Tabernacle 1886 (Beswick possibly assisted Read)   P

BEVAN, John architect
1. SWINDON. Edgware Road. St Paul's. Chancel 1883 (rest of church was Edmund Ferrey 1881). (Demolished 1965)   VCH

BIGWOOD and CO builders of Spa Road, Melksham
1. SEMINGTON. Union Workhouse vagrants' cells 1925-6   RG

BILLINGHAM, Frank George architect
Practice at 15 Market Place, Devizes (Kelly's 1889).

BINNS, H.W architect
1. STANTON FITZWARREN. Stanton House 1935   P

BINYON, Brightwen architect of Ipswich
1. SWINDON. Regents Circus. Town Hall 1889-91, "in a vaguely 17th century Dutch style". P "Danish Renaissance style." JB Originally New Swindon Local Board Offices PW

BIRCH, John architect
Book 'Country Architecture' with cottage designs 1878   JO
1. EVERLEIGH. Manor rebuilding 1882-3, after fire ("probably from designs by John Birch")   VCH
2. GREAT BEDWYN. Estate cottages for Marquis of Ailesbury JO
3. PEWSEY. Cottages at Clench   JO
4. PEWSEY. Curney House   JO

BISHOP & PRITCHETT architects
Charles Bishop born 1855 was senior partner. Practice at Regent Circus, Swindon (Kelly's 1895, 1899); as **BISHOP & FISHER** (Kelly's 1907, 1911)

1. SWINDON. 24A-26A and 27-31 Ashford Road 1910   TS
2. SWINDON. 90-93 Kingshill Road 1911   TS
3. CHISELDON. Holy Cross. Restoration work 1930   WRO

BLACKING, W.H Randall architect
1. WILTON. St Mary & St Nicholas. Restoration work 1933 WRO
2. ODSTOCK. Nunton. St Andrew. Restoration work 1933   WRO
3. DONHEAD ST MARY. St Mary. Restoration work 1934   WRO
4. WEST LAVINGTON. All Saints. Restoration work 1934   WRO
5. DEVIZES. St James. Restoration work 1934, 1939   WRO
6. SEMINGTON. St George. Restoration work 1936   WRO
7. SALISBURY. St Martin. Restoration work 1936   WRO
8. FITTLETON. All Saints. Restoration work 1936   WRO
9. CALNE. St Mary. Restoration work 1936   WRO
10. EDINGTON. St Mary, St Katherine and All Saints. Restoration
    work 1936   WRO
11. NETHERAVON. All Saints. Restoration work 1936   WRO
12. DINTON. St Mary. Restoration work 1936   WRO
13. BROAD TOWN. Christ Church. Restoration work 1937   WRO
14. MERE. St Michael. Restoration work 1937   WRO
15. DOWNTON. St Lawrence. Restoration work 1937   WRO
16. LITTLE CHEVERELL. St Peter. Restoration work 1937   WRO
17. FITTLETON. Parsonage. Repair work 1937   WRO
18. TISBURY. St John. Restoration work 1938   WRO
19. BRADFORD-ON-AVON. Holy Trinity. Restoration work 1938 WRO
20. GREAT CHEVERELL. Parsonage. Repair work 1938-9   WRO
21. DEVIZES. St John. Restoration work 1939   WRO

BLAKE, Robert
1. ORCHESTON. Parsonage 1831   WRO

BLANCHARD, George
1. ROWDE. Parsonage 1817   WRO

BLANDFORD, H and SMITH architects
1. TROWBRIDGE. Union Street. Almshouses 1861   P

BLANTON, John plumber
1. DAUNTSEY. Rectory 1829-33   DR

BLOMFIELD, Sir Arthur William architect 1829-1899
Son of Bishop of London, C.J.Blomfield. NJ Articled to Philip Hardwick.
Architect to the Bank of England 1883. Knighted 1889. DNB High Victorian
Gothic style. NJ
1. CHILTON FOLIAT. Chilton Lodge 1800 (by W Pilkington). But the porte-
   cochere on the east side and an attached wing on the west side are
   attributed to Blomfield.   P
2. LACOCK. St Cyriac. Restoration 1861   P
3. GRITTLETON. St Mary. Restoration 1865-6 ("terribly over-
   restored")   P
4. HULLAVINGTON. St Mary, extensively restored 1871-2, W tower 1880 VCH
                                                         P
5. LUCKINGTON. St Mary and St Ethelbert. Chancel 1872   P
6. COLLINGBOURNE DUCIS. St Andrew. Restoration of church

- 18 -

  (except chancel, by Street) 1876-7   P
7. EAST KNOYLE. St Mary. Restoration 1876   P
8. SWINDON. Edgeware Road, St Paul's Church (but P says by Ferrey)
  1881 JB
9. MARLBOROUGH. College. Museum block 1882-3 (perhaps designed by G.E.
  Street, but completed by his son A.E.Street, with Blomfield) VCH
10. SALISBURY. Cathedral. Restorations, "important" in DNB
11. DONHEAD ST MARY. St Mary. Restoration work 1884   WRO
12. WARMINSTER. St Denys. Restoration 1887-9   JO P
13. WESTBURY. All Saints. Restoration work 1892 (A.W Blomfield
  & Sons)   WRO
14. WARMINSTER. Boreham, St John's Church. Added vestry 1897 JO
15. TIDCOMBE AND FOSBURY. Fosbury, Christ Church. Restoration work 1905 (A.W
  Blomfield & Sons)   WRO

BLORE, Edward architect 1787-1879
Born at Derby. Son of Thomas Blore, topographer and began himself as a
topographical draughtsman. GW Gothic revivalist. CH
1. WILTON. St Mary Magdalene's Hospital 1831   C
2. WARMINSTER. Town Hall 1832   C
3. MARLBOROUGH. College. Domestic buildings 1844 onwards c 1845-50; Chapel
  (Gothic), (now replaced by Bodley & Garner's chapel, 1883-6) P
4. WARMINSTER. 8 evens to 12, Market Place   DOE

BLOUNT, George Leo William architect, surveyor and civil engineer 1870-?
of The Cottage, Winterbourne Earls. Articled pupil to H.Seaver. Practice at
39 High Street, Salisbury 1905 and 1913; at 61 New Street (Kelly's 1915);
at 75 New Street (now **BLOUNT & WILLIAMSON**) (Kelly's 1923); No Williamson in
1927 Kelly's. Photo. and biography in D.
1. SALISBURY. Public library, Young Gallery 1910   P
2. PITTON AND FARLEY. Farley, All Saints. Restoration work 1911 WRO
3. BULFORD. Garrison Church of St George 1920-7 (Blount & Williamson) P
4. SUTTON VENY. Probably St John. Restoration work 1925   WRO

BLOW, Detmar Jellings architect 1867-1929 (or 1939)
Of Hilles, Gloucs. Associated with Gimson at the turn of the century, and
he was Ruskin's last disciple. Linked to the Arts and Crafts movement.
Apprenticed to a mason as well as trained as an architect. Joined SPAB
1890. Pugin travelling scholarship at RIBA in 1892. Partnership with French
architect Fernand Billerey c. 1905. No more country house commissions after
marriage in 1910.   GJS
Specialised in meticulously traditional houses, built of traditional
materials using traditional methods. Estate manager to second Duke of
Westminster, accused 'unjustly it seems' of financial impropriety, sacked
and died in relative disgrace. KP
1. EAST KNOYLE. St Mary's. Helped Philip Webb repair tower 1892-3 GJS
  Burial ground 1899 contains Wyndham Corner, an enclosure with monuments
  etc by Blow.   VCH
2. WOODFORD. Heale House. Restoration and enlargements after 1894. 1910 GJS
  P
3. WILSFORD CUM LAKE. Lake House. Restoration after 1897 and further after
  a fire, 1912   VCH
4. DURNFORD. Little Durnford Manor, lodge late 19thC (possibly by Blow) VCH
5. DURNFORD. Netton, Heale Cottage (cob and thatch) 1900 VCH P (under

Woodford)
6. FONTHILL GIFFORD. Little Ridge (later Fonthill House) 1904 (demolished 1972) P
7. AMESBURY. St Mary and St Melor. Restoration work 1904 WRO
8. AMESBURY. Amesbury Abbey. Alterations to stairwell and long saloon, paving of entrance hall 1904 RCHM
9. WILSFORD CUM LAKE. Wilsford Manor 1904-6 P
10. WEST TISBURY. Newtown, Hatch House. Alterations 1908 P

BLOWE, Hugh mason
The King's master mason at Marlborough.
1. MARLBOROUGH. Castle. Repairs, refitting and new work including a tower for Henry III 1238-9 WAM 85 p72, VCH
2. LUDGERSHALL. Castle. Improvements c1244. Blowe may have been employed there but no evidence. WAM 85

BODLEY, George Frederick and GARNER, Thomas architects
Bodley (1827-1907) was George Gilbert Scott's first pupil. Always Gothic style. Partnership with Garner (1839-1906) began 1869, lasted to 1898. Garner was another of Scott's pupils. FH&P
1. MARLBOROUGH. College Chapel 1883-6 P
2. MARLBOROUGH. College North Block 1893-9 VCH

BOOTH & LEDEBOER architects
1. TISBURY. Wardour Castle. Staircase in SW corner (for Cranborne Chase School) 1960 P

BOTHAMS, Alfred Champney architect 1861-? of Clayton Croft, East Harnham, Salisbury. Father, **John Champeny Bothams**, was City Surveyor of Salisbury 1854-1902. In private practice as engineer and architect at Salisbury from 1883. Municipal work from 1889 and City Surveyor on father's death in 1902. Practice at 39 Castle Street (Kelly's 1895); 9 Endless Street (Kelly's 1899); 32 Chipper Lane (Kelly's 1907 onwards). In partnership with **Bernard Owens Brown** (qv) from 1927. D (photo. and biog.)
1. SALISBURY. Chipper Lane, Public Library 1904 P
2. SALISBURY. Swimming bath, laundry, boot factory, and many houses and shops D
3. CHOLDERTON. Manor House, N service block 1914 VCH

BOTHAMS, BROWN & DIXON architects
i.e D.B.Brown and S.S. Dixon. Practice at 32 Chipper Lane, Salisbury
1. SALISBURY. Solicitor's office, Trethowans, and Jones & Parker RIBA
2. SALISBURY. Pub for Gibbs RIBA
3. SALISBURY. Almshouses - Salisbury City Almshouses RIBA
4. SALISBURY. Sheltered housing schemes for Salisbury District Council RIBA
5. SALISBURY. Extensions for Churchfields Dairies RIBA
6. AMESBURY. Extensions for Churchfields Dairies RIBA
7. DURNFORD. St Andrew. Restoration work 1936 WRO

BOWDEN, F.I architect
WCC County Architect. Architect for Wiltshire Historic Buildings Trust in 1971. R
1. SALISBURY. Bemerton. Secondary Modern School 1956 P

2. CALNE. Wessington. Bentley Grammar School. Main building 1957 P
3. TROWBRIDGE. College of Further Education 1957-9 (with D.H.P.Roberts)
   P
4. SWINDON. County Road. Fire Station 1959   P
5. MARLBOROUGH. Orchard Road. Grammar School 1962   P
6. MARKET LAVINGTON. Comprehensive School 1962   P
7. SWINDON. Police Station 1966-9   P
8. DEVIZES. Sheep Street. Public Library 1966-8   P
9. GRAFTON. Wilton Windmill, repair 1971   R

BOYLE, Richard architect 1694-1753
3rd Earl of Burlington, known as the 'Architect Earl'. Patron of various
Neo-palladians. Brother-in-law of Lord Bruce of Tottenham. RCHM
1. GREAT BEDWYN. Tottenham House. Remodelling 1720s-1730s. The designs were
   made 1721, rebuilding in progress c1730-40. Flitcroft in charge of work.
   Also octagonal summer-house 1743 and banqueting house in the woods
   (demolished 1824).   C RCHM

BRADELL, Darcy and DEANE architects
1. WILSFORD CUM LAKE. Lake House. Plaster ceiling and extensions
   after work of Blow   KP

BRADIE, J
1. ERLESTOKE. High Street, Tilted Lodge c.1800   PW

BRAKSPEAR, Sir Harold architect 1870-1934
Studied with his father, W.H.Brakspear, in Manchester. Practice at The
Priory, Corsham (Kelly's 1895-1927 at least) where he first lived at
Corsham. Then built and lived at Bean Close. Later bought Pickwick Manor
and moved there when knighted. Drawings and papers there with family and
others deposited including clients' ledgers of Sir Harold and his father
back to 1845 (WRO 2512).   PMS (1986)   Photo. and biog. in D   Well known
for restoration of medieval buildings and excavations of monasteries.
Malmesbury Abbey was the first major restoration of his career. Knighted
after the restoration of St George's Chapel, Windsor.   JO
1. SEEND. Cleeve House, alterations 1896   JO
2. HULLAVINGTON. Lych-gate 1897   JO
3. KINGTON LANGLEY. Steinbrook House 1897-8   JO
4. CORSHAM. Pickwick Road, Mansion House (much rebuilt at rear 1897)
   JO
5. WESTWOOD. School extension 1897
6. LACOCK. Bewley Court, restoration 1897 and c1912   WAM 37
7. BRADFORD-ON-AVON. Newtown Brewery. Steel bridge across Wine Street
   1897   Building notice, WRO
8. LANGLEY BURRELL WITHOUT. St Peter. Restoration 1898   P
   Church Tower restored 1929   JO
9. MALMESBURY. Abbey restoration 1899   D
10. CHIPPENHAM. Malmesbury Road, Greenways c.1900   JO
11. LACOCK. Cantax Hill, Raycroft c.1900   JO
12. BRADFORD-ON-AVON. The Hall. Stables 1901-2 and Lodge   P
13. LACOCK. St Cyriac. Remodelling of chancel 1903   P
14. GREAT SOMERFORD. Church porch, restored c1903   VCH
15. CORSHAM. Bean Close (now Wellesley) 124, Priory Street (1904 on glass
    ifo) JO

16. LACOCK. Abbey. Restoration   D
17. ATWORTH. Great Chalfield Manor. Restoration 1905-12   VCH
18. CHIPPENHAM. Rowden Hill, Coters 1907   JO
19. MELKSHAM WITHOUT. Beanacre Manor House restored 1910. Also
    restored Old Manor House   JO
20. LACOCK. Primary School, High Street. Lean-to 1911   JO
21. CHIPPENHAM. St Paul's Church. Stone reredos 1911   SPC
22. WESTBURY. All Saints. Restoration work 1913   WRO
23. SEAGRY. Seagry House. Restored 1914, additions 1915 (burnt down 1949)
    VCH   JO
24. BRADFORD-ON-AVON. Barton Farm, Tithe Barn. Restoration 1914
    VCH
25. STEEPLE ASHTON. Ashton House. Restoration 1920s   VCH
26. BOX. Hazelbury Manor. Reconstruction 1920-5   JO
27. WESTWOOD. St Mary. Restoration work 1923   WRO
28. BRADFORD-ON-AVON. Christ Church. Restoration work 1923   WRO
29. HANNINGTON. Hannington Hall. Restoration work (1924?)   PMS
30. DEVIZES. St Mary. Restoration work 1925, 1931, 1934   WRO
31. ALDBOURNE. St Michael. Restoration work 1925   WRO
32. OGBOURNE ST ANDREW. St Andrew. Restoration work 1925   WRO
33. MANNINGFORD. Manningford Bruce. St Peter. Restoration work 1926 WRO
34. ALLCANNINGS. All Saints. Restoration work 1927   WRO
35. LITTLE SOMERFORD. Hill House 1927   JO
36. MELKSHAM. St Andrew. Restoration work 1929   WRO
37. EDINGTON. Priory. Restoration work 1930   WRO
38. CALNE. St Mary. Restoration work 1930   WRO
39. AVEBURY. St James. Restoration work 1930   WRO
40. KINGTON LANGLEY. Vicarage 1931   JO
41. BISHOP'S CANNINGS. St Mary. Restoration work 1931   WRO
42. PRESHUTE. St George. Restoration work 1931   WRO
43. SWINDON. Christ Church, Cricklade Street (by Sir G.G.Scott
    1851). SE Chapel 1935, known by 1964 as Lady Chapel   VCH
44. CORSHAM. Jaggard's. Restoration   D
45. HEDDINGTON. St Andrew. Restoration work 1938-9   WRO
46. CHIPPENHAM. Greenways. New houses   D
47. CORSHAM. New schools   D
48. SWINDON. Park North, St John the Baptist 1961   VCH
49. BIDDESTONE. St Nicholas Church. W window tracery restoration
    (drawings survive)   JO
50. LACOCK. Porch House and 21 Church Street, restorations   PMS
51. WROUGHTON. Wroughton House?   DOE

BRAKSPEAR, Oswald S architect of Corsham
Son of Sir Harold Brakspear. Practice at Pickwick Manor, Corsham.
Bristol Diocesan Architect with extensive work in churches and parsonage
houses. Various records of work and ledgers for period 1938-1974 deposited,
WRO 2512.
1. COLERNE. One of his first jobs (1930s?) was to Georgianise the water
   tower at Lucknam Park and design the Middle Lodge. JO info from OB
2. CHIPPENHAM. St Paul's Church. Baptistry 1955   SPC
3. WANBOROUGH. Vicarage 1959   VCH

BRAKSPEAR, Sidney architect
Practice at Ash Villa, Corsham (Kelly's 1911)

BRANDON, David architect
(See WYATT and BRANDON)

BRETTINGHAM, Matthew the younger architect 1725-1803
Freeman of Norwich 1769. Not much of his architectural work known. C
1. CHARLTON N. Charlton Park. Alterations to loggia, S and E fronts
   1772-6   P

BREWER, builder
1. STANTON ST QUINTIN. Upper Stanton St Quintin Farm. Work 1852-1854 EG

BRIANT, Henry and **BRIANT, Nathaniel** architects of Reading.
1. NETHERAVON. Parsonage 1838   WRO

BRICK, of the Adelphi, London
1. SUTTON BENGER. Draycot Cerne. Upper Draycott. A brick cottage 1864 (Mr
   Brick won a prize and medal of the Society of Arts for it)   P

BRIDGES, J.B architect
1. OAKSEY. Rectory, alterations 1869   VCH

BRIGGS & GORDON
1. LAVERSTOCK. St Andrew. Restoration work 1906   WRO

BRINKWORTH, Robert E architect
Practice in St Mary Street, Chippenham (Kelly's 1899); at 7 Marshfield Road
(Kelly's 1907). Had a Bath office too.
1. DEVIZES. Bath Road. Grammar School 1906 (now part of St Peter's School)
   P

BROMLEY, Benjamin architect of Corsham
1. NORTH WRAXALL. Parsonage 1815   WRO
2. NETTLETON. Parsonage 1816   WRO

BROMLEY, W.H architect and builder of Corsham
1. MELKSHAM WITHOUT. Shaw. New school and house 1871 (architect W. Smith)
   WRO
2. CORSHAM. Chapel Knap School (plan of site) c1871   WRO
3. CORSHAM. Pickwick Road, Wesleyan Chapel 1878   JO
4. CORSHAM. Town Hall (incorporating parts of Market House of 1784), 1882
   JO

BROOKE, Capt. Joshua Watts architect 1865-?
Born at Burbage. Father (a pupil of McAdam) was a surveyor in Marlborough.
Lived at Rosslyn, Marlborough. Became Surveyor for Marlborough RDC,
Ramsbury RDC etc. D Practice at 12 Kingsbury Street, Marlborough
(Kelly's 1927).
1. PEWSEY. Workhouse (Vagrant Wards addition, 1st prize - 'The Builder' 75
   1898 p584)   JO
2. MARLBOROUGH. Workhouse. Vagrant ward   RG

BROOKS, James architect 1825-1901
Gothic revivalist, usually working in stock brick. Best known churches all
in London.   FH&P

1. MARSTON MAISEY. St James 1874-6   P
2. MARSTON MAISEY. Bleeke House (formerly the vicarage) rebuilt c.1876 P

BROOKS, John painter
1. DAUNTSEY. Rectory 1829-33 (not an artist)   DR

BROWN, builder of Hilperton
1. HOLT. Congregational Chapel 1880 (with Mr Ponton of Warminster)   HH

BROWN, Bernard Owens architect of Salisbury
(See also under A.C.Bothams.)
1. WILTON. South Street. Memorial Hall 1938   P

BROWN, Ford Maddox stained glass designer and painter 1821-93
1. MALMESBURY WITHOUT. Rodbourne. Holy Rood Church. E. Window 1865 (with
   D.G.Rosetti and made by Morris and Co)   VCH

BROWN, Francis builder of Tetbury, Gloucs.
1. MALMESBURY. Westport. School and house 1857   WRO
2. CRUDWELL. Eastcourt. School and house 1858   WRO

BROWN, Lancelot (Capability) landscape gardener 1716-83
1. CHARLTON N. CHarlton Park. Proposed alterations to house, not accepted
   1768   VCH

BROWN, R.J architect
1. CHIPPENHAM. Market Place. A motel by the Angel Hotel 1959   P

BROWNLOW and CHEERS architects of Liverpool
1. SWINDON. Bath Road. Wesleyan Methodist Church 1880   P

BRUNEL, Isambard Kingdom architect and engineer 1806-1859
Born Portsmouth, son of Sir Marc Isambard Brunel, architect and engineer.
Worked in father's office. 1833 appointed engineer to Great Western
Railway. Best known works - Clifton Suspension Bridge designed 1830s and
Paddington Station 1854.   CH
1. BOX. Tunnel 1837-41   P
2. CHIPPENHAM. Station 1841. Brunel's first Italianate work   JO
3. CHIPPENHAM. Marshfield Road, Western Villas (for GWR staff) c1841   SPC
4. SWINDON. Station 1841-2   P
5. SWINDON. Houses, shops, inns, locomotive and wagon servicing and repair
   sheds and foundries - about 300 structures 1840s   RCHM
6. BRADFORD-ON-AVON. Station 1848   JO
7. TROWBRIDGE. Station 1848 (demolished 1984) and goods shed 1848
   (demolished 1985)   WT   JO
8. SALISBURY. Great Western Station 1856 P (closed 1932)   VCH

BRUNSDON, William plasterer of St Peter's parish, Marlborough
Married c.1641, buried June 1662
1. MARLBOROUGH. 132 High Street 1656 (his name and date in plaster) WBR
   Said to have carried out other work at Marlborough.

BRUTON, E of Oxford
1. WESTBURY. Cemetery Chapels 1857   JO

BUCKLER, John architect and topographical artist 1770-1851 FSA
Said to have been articled to C.T.Cricklow, a Southwark architect. GW
Commissioned by Sir Richard Colt Hoare to make 700 watercolour drawings of
churches, great houses and ancient buildings of Wiltshire 1803-1810.
1. STOURTON WITH GASPER. Stourton. Inn. Designed new front in 1812 for Sir
   Richard Colt Hoare. Never executed. (Watercolour at Devizes Museum
   library)

BURFORD, James architect
1. DINTON. Hydes House, restoration of dovecote for National Trust 1951-3
2. POTTERNE. Porch House. Repair 1957

BURGESS, James stone mason, architect and builder of Westbury
At Fore Street in 1830 and Brook Street in 1842 ETD Said to be
conscientious, loved his work, (re Holt Church).
1. WEST ASHTON. School and house 1846   WRO
2. HEYWOOD. Parsonage 1848   WRO
3. EDINGTON. Priory Church, restoration 1887 (C.E.Ponting architect)
4. WESTBURY. Church Street, Public Baths 1887-8   WT
5. HOLT. St Katherine's Church, alterations 1891-4 (C.E.Ponting
   architect).

BURLINGTON, Earl of
See under BOYLE, Richard

BURLISON and GRYLLS
1. TROWBRIDGE. St James's Church, window 1874   JO

BURN, William architect 1789-1870
Son of an Edinburgh architect and builder. One of Scotland's leading
architects. His later work showed great freedom of style. GW
1. FONTHILL GIFFORD. A mansion for the Marquess of Westminster
   1846-52 (demolished 1955)
2. BROMHAM. Bowden Hill, Spye Park. New house 1863-8 built for J.W.G
   Spicer in gabled Elizabethan/Francois Ier style, burnt 1973 and
   demolished 1975-85. Nothing left but entrance porch and part of service
   wing basement. Present house is former stables said to date from 1654
   with c1870 extension   P   JO

BURTON, Decimus architect 1800-1881
Son of James Burton, a speculative London builder. Designed arch at Hyde
Park Corner. CH GW Work at Bath?
1. CALNE. Quemerford, The Croft (former Moon Croft), attributed
   to Burton (Crisp and Cowley advert)

BUTCHER, R & SON builders of 39 George Street, Warminster
Firm founded about 1800 by Robert Butcher, carpenter. Had cousins who were
stone masons and the family had been brickmakers at Crockerton going back
to 1743. Two further Robert Butchers ran the firm and then Frederick
Butcher took over from his uncle in 1919. Son Geoffrey Butcher with Matthew
Butcher (sixth generation) running in 1991.   GB
1. WARMINSTER. Boreham Road, Prestbury House pre-1919   GB
2. WARMINSTER. Brick Hill, The Croft pre-1919   GB
3. WARMINSTER. Hospital 1918-1939 period   GB

4. WARMINSTER. St Boniface, library 1918-1939 period   GB
5. HORNINGSHAM. Village Memorial Hall 1918-1939 period (mostly from
   materials from old Reformatory School, Cannimore)   GB
6. WARMINSTER. New Close Primary School c1953   GB
7. MERE. Old People's Home. After 1945   GB
8. BOYTON. Broadleaze House 1964   GB
9. SALISBURY. The Close, Cathedral School Principal's house c1966-7 GB
10. CORSHAM. School 1982?   GB
11. HINDON. Vicarage. After 1945   GB
12. TEFFONT. Vicarage. After 1945   GB
13. DONHEAD ST MARY. Vicarage. After 1945   GB
14. NETTLETON. Nettleton Mill, conversion to house, after 1980   GB
15. STEEPLE LANGFORD. Bathampton. Restoration of medieval barn   GB

Post 1945 work also includes restoration work at Lacock village (for NT), Longleat House, Stourhead House, Sutton Veny House, Boyton Manor, Stockton House, Bapton Manor, Fisherton de la Mere House and Bathampton House.

BUTTERFIELD, William architect 1814-1900 of 4 Adam Street, Adelphi, London. Born London. High church Gothic revivalist. Earliest church - Coalpit Heath 1844.

1. OGBOURNE ST ANDREW. St Andrew. Restoration 1847-9   P
2. OGBOURNE ST ANDREW. Vicarage 1848 (now Tresco House)   P
3. CHIRTON. St John's Church restored 1850 (VCH gives restorer as James
   Dutch ?builder)   JO
4. AMESBURY. St Mary and St Melor. Restoration 1852-3 (including
   new pulpit)   VCH   P
5. SALISBURY. St Nicholas Hospital. Restoration 1854   P
6. DOWNTON. Charlton. School 1857-8 (now a private house)   P
7. ALDBOURNE. School 1858 (demolished)   WRO   P
8. LANDFORD. St Andrew 1858 (except N doorway)   P
9. LATTON. St John the Baptist. Chancel 1858-63 (1861)   P
10. DOWNTON. Standlynch. Trafalgar House. Balustrade in front of house 1859.
    Church restoration 1859-66   P
11. AMESBURY. Cemetery Chapel 1860 (Kelly)   JO
12. CLYFFE PYPARD. St Peter. Restoration of chancel 1860, rest of church
    1873-4   P
13. DOWNTON. Charlton. Vicarage 1860-2   P
14. CASTLE EATON. St Mary. Restoration 1861-3   P
15. BREMHILL. St Martin. West window 1862-3. Pulpit?   P
16. LYNEHAM. St Michael. Chancel 1862-5   P
17. ALDBOURNE. St Michael. Restoration 1863-7   P
18. BLUNSDON ST ANDREW. St Andrew 1864-8   P
19. HEYTESBURY. St Peter and St Paul. Restoration 1865-7   P
20. HILMARTON. Highway. St Peter 1866-7 (now converted to a house) P
21. BLUNSDON ST ANDREW. Broad Blunsdon. St Leonard 1870 (except C13 S.
    arcade)   P
22. WHITEPARISH. All Saints exterior 1870   P
23. LANDFORD. Vicarage altered or rebuilt c1871   WRO   JO
24. PURTON. St Mary. Restoration 1872   P
25. WEST KNOYLE. St Mary, some work in 1873 (i.e before Allen's
    1878 church)   WRO
26. DINTON. St Mary. Restoration 1873-5   P
27. SALISBURY. West Harnham, St George. Restoration 1873-4 (inc. rebuilding
    of E end of chancel, W wall of nave and nave roof; the N porch was

turned into a vestry and a new porch built on to the S side of the nave) RCHM
28. KNOOK. St Margaret. Restoration 1874-6   P
29. DINTON. School 1875 - plans WRO   JO
30. ASHTON KEYNES. Holy Cross. Restoration 1876-7   P
31. NETHERHAMPTON. St Catherine 1876-7 (except C18 tower)   P
32. BREMHILL. Foxham. St John the Baptist 1878-81   P
33. WINTERBOURNE MONKTON. St Mary Magdalene 1878 (except early C14 south porch, north doorway, C13 lancet in chancel south window & C16-17 nave N window)   P
34. DINTON. Baverstock. St Edith. Restoration 1880-2 (inc new screen) P
35. SALISBURY. Theological College chapel 1881   JO

BUTTON, builder of Calne
Possibly John Button, builder 1793-8, Thomas Button, mason 1793-8 or Edward Button of Church Street, carpenter and joiner in 1793-8 and 1822   ETD
1. AVEBURY. St James. Tuscan columns in arcade 1812   P

BUTTON, John builder and mason of Calne
1. CALNE WITHOUT. Bowood. Work there 1755 when he was described as 'a little backward' by Lord Shelburne. Also in 1762 and in 1769-73 (when principal mason) under Adam brothers. WAM 41, 1922, p514 et seq.

CACHEMAILLE DAY, N.F architect
1. SALISBURY. North Bemerton. St Michael 1956-7   P

CAMPBELL, Colen architect 1673-1729
Introduced Palladianism to this country. First volume of his Vitruvius Britannicus published 1715. Other buildings: Baldersbury Park, Yks 1720-1; Burlington House, London 1718-9; Mereworth Castle 1722-5; Compton Place, Eastbourne 1726-7 FH&P
1. STOURTON with GASPER. Stourhead House 1721-4, based on Villa Emo at Fanzolo. His portico was not built till 1841   P RCHM

CAMPBELL, W Hinton architect
1. BROAD TOWN. Christ Church 1846   P

CANNON, James architect or builder
1. LIDDINGTON. School 1875   WRO

CARD, John carpenter and builder of Westbury
1. WESTBURY. Old workhouse, alterations 1837 (architect Evans) RG
2. SALISBURY. Fisherton Anger. School and house 1844

CARLINI, sculptor
1. CALNE WITHOUT. Bowood House. Marble sarcophagus in Mausoleum 1765 WAM 41, 1922, p513

CAROE, W.D architect (CAROE & PASSMORE)
1. SALISBURY. Stratford-sub-Castle. St Lawrence. Restoration 1904-5 VCH
2. SALISBURY. St Martin. Restoration work 1905   WRO
3. LAVERSTOCK. St Andrew. Restoration work 1907   WRO
4. PITTON AND FARLEY. Farley, All Saints. Restoration work 1909 WRO
5. ODSTOCK. Nunton. St Andrew. Restoration work 1920   WRO

6. MERE. St Michael. Restoration work 1931 (with W.H.Blacking) WRO
7. CALNE. St Mary. Restoration work 1934   WRO

CARPENTER, R. Herbert and INGELOW, Benjamin
1. STEEPLE LANGFORD. All Saints. Restoration work 1875   WRO
2. EAST KNOYLE. Knoyle House. Rebuilding 1880. Engraving-'The Builder'
   VCH

CARPENTER, Richard Cromwell architect 1812-1855
1. DEVIZES. St Mary. Early Victorian restoration and east window P
2. DEVIZES. St John's vicarage   WRO   JO

CARRINGTON, Dora
1. LUDGERSHALL. Biddesden House. Trompe l'oeil scenes painted in several
   windows (with Roland Pym) 1930s   VCH

CARSON AND MILLER iron founders of East Street, Warminster
Also making ironwork for county bridges c1829-30. PL
1. WARMINSTER. Metalwork for workhouse 1836-7   RG

CARTER, Benjamin and Thomas stone masons
1. CALNE WITHOUT. Bowood House. Marble chimney pieces to design of Robert
   Adam 1763   WAM 41

CARTER, Owen B architect of Winchester
63 of his drawings of Wiltshire churches and Porch House, Potterne dated
between 1847 and 1850 are at Devizes Museum.   PMS
1. SALISBURY. Poultry Cross. Designs of upper part 1834, built
   by W Osmond 1852-4   RCHM

CASSEY, Thomas architect
Practice at Milford Street, Salisbury (Kelly's 1875, 1880)

CAUS, Isaac de mathematician fl. 1644
Nephew of the engineer and architect Solomon de Caus. Assistant to Inigo
Jones. They were Gascons, and the name is often spelled Caux. (See 'Wilton
House and English Palladianism' by RCHM).
1. RAMSBURY. Manor House. Stables mid C17. P (Attribution by analogy with
   stables at Wilton House)   VCH
2. WILTON. Wilton House. Laid out the gardens 1632-5. Stable court
   rebuilt c1632-5. S front rebuilt 1636. Palatial design, eventually
   completed in reduced form. Possibly advised by Inigo Jones.
   P RCHM DNB VCH

CHAMBERS, Sir William architect 1723-1796
Studied in Paris under J-F Blondel 1749 & in Italy 1750-5. Appointed
architectural tutor to the Prince of wales, 1756. Became, with Robert Adam,
Architect to the King, 1760, Comptroller, 1769, and Surveyor-General, 1782.
Style based on English Palladianism, veering to neo-Classicism.   F,H & P
1. WILTON Wilton House. An archway, c1759, originally on top of the
   hill to the S of the house, but brought down by Wyatt, (early C19)
   to close the forecourt. P 'the first work of stone I executed in
   England'   RCHM
2. WILTON. Wilton House, Casino, on the hill in front of south facade,

    c1759   ized RCHM P
3. WILTON. Wilton House. Remodelling of W range including new library 1762 RCHM
4. AMESBURY. Amesbury Abbey. The Chinese temple in the grounds 1772.   P
Possibly only embellishment of existing building   RCHM

CHAMPION, W Scott architect
1. BISHOPSTROW. St Aldhelm. "Victorianized" by Champion 1867-7   P

CHANDLER, builder and timber dealer of Pewsey
1. PEWSEY. Workhouse 1836 (Cooper architect)   RG

CHANTREY,
1. LONGBRIDGE DEVERILL Church, monument to Marquis of Bath d 1837   JO

CHAPMAN, John mason
In the King's works from c1541.
1. HORNINGSHAM. Longleat. Extension and rebuilding in the 1550s & 1560s P
2. LACOCK. Lacock Abbey, ascription of stone table in tower c1550   NT

CHAPMAN, Joseph of Frome, Somerset
1. MAIDEN BRADLEY. National School 1845   WRO

CHAPMAN & SONS (of Frome, Somerset?)
1. LONGBRIDGE DEVERILL. Hill Deverill, The Assumption 1843 (a rebuilding?) P
2. IMBER. Parsonage 1844 (demolished)   WRO

CHESTERTON, Maurice architect
1. SEAGRY. Upper Seagry. Hungerdown House 1914   VCH   JO

CHIVERS, W.E builder of Devizes
William Chivers, carpenter and builder of Sidmouth Street 1842.   ETD
1. ROUNDWAY. County Mental Hospital. The Annexe, begun 1913   RKN

CHRISTIAN, Ewan architect of London 1814-1895
A Gothic revival architect.
1. BLUNSDON ST ANDREW. Broad Blunsdon Parsonage (probable)   P
2. FIGHELDEAN. St Michael and All Angels. Chancel etc restored 1858-9 VCH
3. SEMINGTON. St George 1860   WRO
4. WESTBURY. All Saints. Restoration work 1862   WRO
5. MARKET LAVINGTON. St Mary. Restoration 1862 (1864 VCH)   P
6. CRICKLADE. St Sampson. Restoration of S aisle wall and windows 1864 P
7. MARKET LAVINGTON. Manor House 1865   P
8. POTTERNE. St Mary. Restoration work 1871   WRO
9. WINTERBOURNE MONKTON. Vicarage altered or rebuilt 1873   JO
10. PITTON AND FARLEY. Farley. All Saints. Restoration 1875   P
11. EASTERTON. St Barnabas' Church 1875 by 'Mr Christian'   SJ
12. POTTERNE. Porch House. Restoration 1876   VCH
13. PITTON AND FARLEY. St Peter. The Victorian parts of the church 1880   P
14. SALISBURY. Deanery. Restoration 1881   WRO
15. SALISBURY. St Edmunds parsonage. Restoration 1881   WRO

16. BISHOPSTONE N. St Mary. Restoration 1882 VCH
17. TISBURY. St John the Baptist. Reredos, presented 1884, designed
    by Christian. P
18. SHERSTON. Holy Cross. East window and chancel, restoration
    ("drastically restored") P
19. FROXFIELD. All Saints. Restoration 1891 WRO

CHRISTIAN, Henry architect of London
1. SHREWTON. Parsonage 1876 WRO
2. SHREWTON. Maddington, Vicarage WRO JO

CHRISTOPHER, John Thomas architect of London (of CHRISTOPHER & WHITE)
1. CHILTON FOLIAT. Parsonage 1891 WRO

CIPRIANI, G.B painter
1. CALNE WITHOUT. Bowood House. Paintings in Robert Adams' decorative
   panels 1761-73 P WAM 41

CLACEY, N.E architect
1. MANNINGFORD. Manningford Bohune. All Saints 1859 P

CLARKE, John joiner
1. SEMINGTON. Whaddon House. Work 1682-3 LP

CLARKE, Samuel architect, engineer and surveyor of Salisbury
1. SALISBURY. West Harnham. National School 1863 WRO

CLARKE, William architect of Bruton, Somerset
1. FOVANT. Parsonage 1797 WRO

CLAYTON AND BELL glaziers
1. CHAPMANSLADE. Church, E window 1866-7 JO
2. SUTTON VENY. St John's Church, stained glass JO

CLERMONT, Andien de painter
1. WILTON. Wilton House. Ceiling and staircase paintings c1739 RCHM

CLUTTON, Henry architect 1819-93
Distinguished architect. JO
1. GRITTLETON. Grittleton House. Original design 1842, begun 1848 (c1854
   JO), but completed by James Thompson. P
2. BRADFORD-ON-AVON. Frankleigh House. Rebuilt on old core 1848 (builders
   Jones Bros) JO
3. STEEPLE ASHTON. St Mary. Chancel rebuilt 1853. Pulpit also by him.
   P
4. SALISBURY. Cathedral. Chapter House, restoration 1856, by J.B.Philip,
   under Clutton. P

COCKERELL, Charles Robert architect 1788-1863
Studied under his father S. P. Cockerell, and assisted Sir Robert Smirke.
Cambridge University (Law) Library, 1836-42; Ashmolean Building, Oxford,
1841-5. Professor of Architecture at the Royal Academy. F, H & P
A master of the most austere phase of the Greek Revival. GW
1. CALNE WITHOUT. Bowood. Chapel 1821. Library in the Adam range. P

2. MALMESBURY. Burton Hill House 1840 or 1842 - destroyed by fire 1846.
Cockerell, the owner's brother, also designed its replacement? P VCH

COCKERELL, Samuel Pepys architect 1754-1827
Best remembered for his Indian-style building, Sezincote, Gloucs (1803)
F, H & P
1. SALISBURY. St Edmund's College (now the Council House), N wing
1788-1790   RCHM

COE & GOODWIN architects
1. LEA AND CLEVERTON. Garsdon. All Saints, rebuilt 1856   VCH

COLBORNE, Arthur Joseph builder of County Works, County Road, Swindon
1. SWINDON. 24A-26A and 27-31 Ashford Road 1910   TS
2. SWINDON. 90-93 Kingshill Road 1911   TS
Both the above with Bishop and Fisher architects.

COLE, Eric design group (later COLE, Eric and PARTNERS)
Office at 12 Bath Road, Swindon (head office Cirencester).
1. SWINDON. Penhill Drive. Penhill Free Church 1959   VCH
2. MARLBOROUGH. York Place. Old people's dwellings 1972   P

COLE, J.J
Architects, Engineers & Building Trades Directory 1868
1. AMESBURY. Amesbury Abbey. Alterations before 1853 and in 1885
RCHM JO

COLLINS, P.G builder of Southview, Freshford, Som.
1. TROWBRIDGE. Frome Road. Gospel Hall 1924   WRO

COLSON, John architect
1. COLLINGBOURNE KINGSTON. St Mary. Restoration 1862   P

COMBES, Cyrus architect of Tisbury
1. SUTTON MANDEVILLE. Parsonage 1869   WRO

COMPER, Sir J. Ninian architect and designer 1864-1960
A church architect with an important practice from the 1890s.
In particular the crypt of Street's St Mary Magdalene, Paddington.
1. DEVIZES. War memorial in St. Peter's churchyard   VCH

CONRADI, William architect
Practice at Vicarage Street, Warminster (Kelly's 1880)

COOMBS, William architect
1. BROAD CHALKE. National School, addition to schoolroom 1872 WRO
2. BOWER CHALKE. School house and additional classroom 1875   WRO

COOPER, William auctioneer and appraiser of Bell Street, Henley-on-Thames
1. PEWSEY. Workhouse 1836   RC
2. MARLBOROUGH. Workhouse 1837 (children's convalescent home after
1929)   RG VCH

COPPER, David carpenter
1. DAUNTSEY. Rectory 1829-33   DR

CORFIELD, Charles architect
Practice at Battlesford, North Newnton (Kelly's 1875)

COMLEY, G.J.D architect
1. CALNE WITHOUT. Derry Hill, Christ Church. Restoration work 1934
   WRO

CREEKE, Christopher Crabbe architect of Bournemouth 1820-1886
Surveyor to and Inspector of Nuisances at Bournemouth. RG
Designed several buildings in Dorset including workhouse at Blandford
Forum.   BG per RG
1. CHIPPENHAM. Rowden Hill. Workhouse (now hospital) 1859   RG
2. TISBURY. Workhouse 1869   RG

CRICKMAY & SON of Weymouth 1830-1907 (G.R.Crickmay)   JO
1. SALISBURY. St Nicholas' Hospital. Restoration of S building 1884
   RCHM
2. SALISBURY. St Martin. Restoration 1885   RCHM
3. SALISBURY. Crane Street, Church House. Alterations 1887   P
4. MARLBOROUGH. Bear Inn late C19   P
5. SALISBURY. South Canonry. Restoration work 1890   WRO

CRIPPS, W.H architect
1. SWINDON. Queen's Drive. Whitbourne Avenue. Methodist Church 1959 P

CRITZ, Emanuel de painter
1. WILTON. Wilton House. Wall and ceiling paintings c1647-54 RCHM

CROCKER, E.H architect
1. MARLBOROUGH. College. B House 1880s   P

CROOK, T builder of Whiteparish
1. WINTERBOURNE STOKE. Vicarage alteration or rebuilding 1851 WRO JO

CROOK, William builder
1. DINTON. National School 1875 (architect W.Butterfield)   WRO

CUNDY, Thomas the elder builder & architect 1765-1825
Apprenticed to builder in Plymouth. GW At 28 Clerk of Works to S.P.
Cockerell, later set up on his own as architect and builder in Pimlico. C
1. GREAT BEDWYN. Tottenham House. Stables 1818. Extensive enlargements and
   sumptuous remodelling of Palladian house 1825 P RCHM

CUNDY, Thomas the younger   architect of London 1790-1867
Succeeded to his father's practice on the latter's death in 1825. C
1. LITTLE CHEVERELL. St Peter. Rebuilding 1850   VCH
2. MARKET LAVINGTON. Parsonage 1851 alt. or rbld.   JO   WRO
3. PEWSEY. St John the Baptist. Restoration work 1853   WRO
4. BULKINGTON. Christ Church 1860   P

CURNIE, W.W architect of Corsham
1. MELKSHAM. St Andrew. Restoration work 1936   WRO

CURTIS, Henry architect of Salisbury
1. AMESBURY. Parsonage 1859 (1851 JO)   WRO

CURTIS, W.R.H architect of Trowbridge
1. HEYWOOD. Holy Trinity. Restoration work 1936   WRO

DANCE, George junior architect, 1741-1825
Son of George Dance (d 1768), architect of the Mansion House, London (1739-52). Jnr studied in Italy, winning gold medal at Parma in 1763 with advanced neo-classical designs. His first building after return from Italy was All Hallows, London Wall, 1765-7. Designed Theatre Royal, Bath c1806. NJ Founder member of Royal Academy, 1768. His pupil was Soane. F,H & P
1. CALNE WITHOUT. Bowood. A gallery c1790   P

DANIELL, Thomas R.A 1749-1840
Acknowledged expert on Indian architecture. GW Designed Hindoo Temple at Melchet Park (then Wilts now in Hants) in late 18C

DARCY, BRADDELL & DEANE architects
1. WILSFORD CUM LAKE. Lake House. Rebuilding after the fire of 1912 P
2. WOODFORD. Upper Woodford. A farmhouse 1935   P

DARLEY & WILKINSON architects
George Wilkinson and John Darley.
1. MALMESBURY. Workhouse. Plan but not accepted c1838   RG

DARLEY, James architect and builder   -1821
His tomb is in Hullavington churchyard. DOE   Will proved 12.7.1822. Left architectural drawings etc. to son John. RH
1. CHARLTON N. Charlton House. Clerk of Works during 1772-6 alterations. C
2. MALMESBURY. Abbey. Submitted estimate for repairs 1788, not known if
   carried out.   RH

DARLEY, John architect and builder of Chippenham
Son of James Darley. RG At High Street in 1842. ETD
1. BOX. Church of St Thomas A'Becket. Alterations 1824.   JO

DARLEY, John & SONS architects of Chippenham
1. CHIPPENHAM. St Mary Street school and church yard 1857   WRO
2. KINGTON LANGLEY. School 1857   WRO
3. BRINKWORTH. School 1869   WRO

DARLEY, Richard architect
Son of John Darley. RG Practice (with John DARLEY, brother) at High Street, Chippenham (Kelly's 1880, 1895): then at 7 Market Place, Chippenham.
1. LITTLE SOMERFORD. School and house 1872   WRO
2. BRINKWORTH. School. Plans (alterations?) 1872   WRO
3. CHIPPENHAM. St Andrew's nave, chancel, N aisle 1875-8   P

DAVIS, Charles Edward architect of Bath   Died 1902 aged 75.
Inherited practice from his father, Charles Davis. In April 1862 became
Surveyor of Works for Bath. Many felt it was unethical to continue with
private practice as he did. Founder and Hon. Sec of Bath School of Art.
Discovered and excavated Roman Baths at Bath. BC Memorable for his last
work there, the Empire Hotel. JO Range of church work - restorations, minor
additions and some new buildings, usually Gothic in style. A number of
country houses. Used Italianate Norman stle in early 1860s. BC

1. TROWBRIDGE. The Down Cemetery buildings 1855   JO
2. TROWBRIDGE. Studley St John's. School and house 1855 WRO
3. BROMHAM. Parsonage 1858   WRO
4. TROWBRIDGE. Market House (next to Town Hall) 1861   P
5. TROWBRIDGE. Pulpit of Holy Trinity Church 1861   JO
6. CALNE. Silver Street, Woodlands   JO
7. MONKTON FARLEIGH. Monkton Farleigh House (possible)   JO

DAVISON, Thomas R. architect
Practice at 3 North Street, Swindon (Kelly's 1880) Little known apart from
Victoria Institute except for winning another competition, Torquay Town
Hall 1911.   JO
1. TROWBRIDGE, Castle Street, Victoria Technical Institute 1897 P
   Won competition, judge E.W.Mountford, architect, of the Old Bailey.
   JO Demolished c1981

Facade of the Victoria
Institute in 1981 (WBR
collection)

- 34 -

DAWBER, Sir (Edward) Guy architect 1861-1938
Practised in the Cotswolds, and from 1891 in London. Mainly a country house architect. Knighted 1936. DNB
1. WINSLEY. Conkwell Grange 1907   P
2. REDLYNCH. Hamptworth Lodge 1910-12   P

DAWKINS, Thomas builder of Wilton
1. WILTON. North Street. 36 cottage houses, house and shop c.1900 HHW

DAY, Charles carpenter
1. DAUNTSEY. Rectory 1829-33   DR

DEANE, John
1. MARLBOROUGH. House (new house commissioned 1684)   VCH

DELAMOTTE, Vallin architect
1. WILTON. Wilton House. Riding House. Elegant French classical designs 1755 but built to a simpler design   RCHM

DENHAM, G architect
Practice at 14 Brown Street, Salisbury (Kelly's 1911)

DEREHAM, Elias de Canon of Sarum and Wells   Died 1245
In charge of the King's Works at Clarendon Palace and Winchester Cathedral.
1. CLARENDON PARK. Clarendon Palace, second stage of building 1240
2. SALISBURY. Cathedral, possible designer. P pleads his case, while others make a case for Nicholas of Ely.
3. SALISBURY. Close. Leadenhall (Leyden) House. One of the earliest houses in the Close, built as a pattern for others.   VCH

DEVALL, John stone mason
1. WILTON. Wilton House. Palladian Bridge 1736-7   RCHM

DEVERELL, Edward rough mason of Bradford-on-Avon
1. BRADFORD-ON-AVON. Middle Rank. House. Building lease from A. Methuen on site 1698   MP

DEVERELL, Edward and John
1. BRADFORD-ON-AVON. St Margaret's Street, Congregational Chapel 1740 RCHM

DEVEY, G architect
1. FONTHILL GIFFORD. No. 74 'Good early example of Vernacular Revival ... the style strongly suggests ... by G. Devey' 1862   DOE

DINSLEY, W.H architect
1. SALISBURY. Milford Street, Elim Church 1896 P, closed 1958   VCH

DOWNING & RUDMAN builders of Chippenham
William Downing, farmer and builder early 19C. Son, Simon, carried on building business at Old Road premises from 1830. Simon Downing, stone mason at Timber Street in 1842. ETD. Joined by grandson A. Rudman in late 19thC. Known for quality of wood carving, work on church buildings, including Worcester Cathedral and Battle Abbey. Kelly 1931 says

'specialists in oakwork and medieval ironwork and the complete renovation of ancient buildings'. In 1936 firm bought by Edward Bent. **DOWNING, RUDMAN & BENT** built RAF and military establishments. Using this title in 1965. HQ in 1987 at Spanbourne Avenue. CC
1. CHIPPENHAM. St Paul's Church. Choir vestry extension and organ chamber. Altar rail converted to pew fronts 1911 SPC
2. CHIPPENHAM. Streets including Canterbury Street and Downing Street CC
3. CHIPPENHAM. Greenway Park estate and other sites S of Malmesbury Road. Plans 1914 SPC CC
4. CHIPPENHAM. Bridge Centre and NAAFI Hall (now Police Station) 1950s CC

DRAKE & PIZEY architects of Bristol
1. SWINDON. Swindon Empire (originally Queen's Theatre), from designs by Drake & Pizey, Flemish Renaissance style 1897 JB

DREW, Edward architect 1867-? of 93 Victoria Road, Swindon
Born Highworth. Son of William Drew (qv). Senior partner of Drew & Sons, Swindon who were at 28 Regents Circus in 1923 (Kelly). Photo. and biography in D.
1. SWINDON. Board Schools at Clarence Street, Gorse Hill, Rodbourne Road, Rodbourne Cheyney. D
2. HAYDON WICK. Board School D
3. STRATTON ST MARGARET. Board School (Beechcroft 1892? PMS) D
4. SWINDON. Queen Victoria Hospital. Memorial Lodge D
5. SWINDON. Conservative Club D
6. SWINDON. Gorse Hill Baptist Church JO
7. SWINDON. Queen Victoria Memorial, Conservative Club c1901 JO
8. SWINDON. Fleetwood Road, Foresters Arms JO
9. SWINDON. Beechcroft, late 19thC JO
10. SWINDON. Goddard Avenue, house for Mr Wilmer and house for Mr S.Smith JO

DREW, G. L architect of Swindon (Known as ARCHITECT DREW)
Practice at Oxford House 57 Victoria Road, Swindon.

DREW, William architect of Swindon
Practice at 4 North Street, Swindon (Kelly 1880-1895); 22 Victoria Street (Kelly 1899); and 28 Regent Circus (Kelly 1907 onwards).
1. STRATTON ST MARGARET. School and house 1870 WRO
2. SWINDON. Rodbourne Cheney. Haydon School 1874 WRO
3. SWINDON. 8-21 Exmouth Street, early 1880s TS
4. SWINDON. 56-64 Exmouth Street, early 1880s TS
5. SWINDON. 1-19 William Street, early 1880s TS
Entries 3-5 built by Edwin Harvey (qv).

DUTCH, Charles builder of Great Cheverell
Also sub-postmaster in 1899 GL

DUTCH, James architect
1. CHIRTON. St John the Baptist. Restoration 1850 (see Butterfield) P

DUTCH, Joseph builder of Great Cheverell
At least from 1883-1899 GL

DYER, Charles architect of Bristol
1. ALDBOURNE. Parsonage 1833   JO   WRO

DYER, John architect, millwright and engineer
At Market Place, Trowbridge in 1830. ETD
1. TROWBRIDGE. Manvers Street, Methodist Chapel 1836 (demolished early 1970s). P   John Dyer was a member of the congregation VCH

Methodist Chapel,
Manvers Street,
Trowbridge
(WCC Planning Dept.)

DYER, William architect of Alton
1. PATNEY. Rectory 1833   VCH

DYKE, D.N architect
1. SALISBURY. Chipper Lane, Telephone Exchange 1936   P

EASTON & ROBERTSON architects
1. LUCKINGTON. Luckington Court (of early 18C), altered 1921   WWA JO

EDEN, F.C architect
1. GREAT SOMERFORD. Church. Designs for painting chancel ceiling 1901 VCH
2. WYLYE. Fisherton Delamere. St Nicholas. Restoration work 1912 WRO
3. CODFORD. St Peter. Restoration work 1912   WRO
4. LONGBRIDGE DEVERILL. SS Peter & Paul. Restoration work 1919   WRO

EDIS, R.W architect of London
1. GREAT BEDWYN. St Katharine's Church. Parsonage   WRO

EDWARDS, Edward architect of Warminster
1. NORTON BAVANT. All Saints. Restoration work 1869   WRO

EDWARDS, George carpenter and joiner of Warminster
At Silver Street in 1830. ETD
1. WARMINSTER. Boreham Road. Re-erection of turnpike house of 1828 removed from East Street 1840   HA

EDWARDS, J.A stained glass maker of Oxford
1. ORCHESTON. Church of St George, E & W windows (Gentleman's Mag. 1833 2.458)   JO

EDWARDS, J Ralph architect of Bristol
1. CHIPPENHAM. St Paul's Church. Peace Memorial, dedicated 1951   SPC

EDWARDS P.W & WEBSTER architects of Chippenham
Edwards joined W. Rudman (qv) as an office boy in 1923. Studied in spare time and qualified 1936. Became partner to Rudman in 1938. Continued practice alone when Rudman died in 1939 and joined by Webster in 1948. Practice was one of those forming the Wyvern Design Group in 1952. PE

EGGINTON, Maria window designer of Birmingham
1. WILTON. Wilton House. Cloister windows 1806-7 (recommended by Wyatt) RCHM

EGINTON HARVEY, architect of Worcester 1809-49
County Surveyor for Worcester.
1. HEYWOOD. Heywood School 1837. Neo-Jacobean   JO
2. WESTBURY. Bratton Road, School 1847   JO
3. HEYWOOD. Holy Trinity Church 1849   JO

ELKINS, E.J architect
1. SALISBURY. Winchester Street, Co-op 1926   P

ELLIOTT, Samuel architect of Newbury, Berks.
1. BUTTERMERE. Parsonage 1876   WRO

ELLIOTT, Thomas builder of Wilton
1. WILTON. Shaftesbury Road, 26 evens to 34 Carvillium Terrace 1892 HHW
2. WILTON. Shaftesbury Road, Dalston Terrace 1896   HHW

ELLIS, Sir Clough Williams architect
Well-known for Portmeirion estate, Wales.
1. WILCOT. Oare, Oare House. Additions of two wings 1921, 1925 P
2. WILCOT. Oare, Cold Blow 1922   P
3. WILCOT. Oare. A terrace of cottages at S end of village   P
4. WILCOT. Oare, school   VCH

ELMBANK FOUNDRY ironfounders of Glasgow
1. SWINDON. Quarry Road. Bandstand in Town Gardens   late 19thC

ELMS, William thatcher
1. STANTON ST QUINTIN. Upper Stanton Farm, repairs 1836   EG

ELY, Nicholas of
1. SALISBURY. Cathedral, possibly the designer of the cathedral   P

ERLESTOKE, Thomas
1. SALISBURY. Old Sarum Castle. Clerk of the Works for repairs 1366 RCHM

EVANS, T.L architect of 2 Little Stanhope Street, Mayfair, London
Evans had by 1837 been Clerk of the Works for Kempthorne (q.v.) but was no
a qualified architect. RG
1. WESTBURY. Old workhouse, alterations 1837   RG
2. WILTON. Workhouse. His plan adopted by Guardians but refused by London
   Inspector due to his lack of qualifications. Hunt was accepted.

FARLEIGH, Richard of mason
1. SALISBURY. Cathedral. Spire 1334   P

FAYREBOWE, John carpenter of Bishopstrow, fl. 1444
1. SALISBURY. Blue Boar Hall, rear of 41 Market Place. Contract 1444
   (employed by Wm. Ludlow of Hill Deverill)   WAM 15

FERREY, Benjamin architect 1810-80
Hon. diocesan architect, Wells, 1841-1880.   DNB
1. GRAFTON. St Nicholas 1842-4   JO
2. CHILTON FOLIAT. St Mary. Restoration 1845 & 1865   P & WRO
3. STOCKTON. Stockton House restoration 1877-1882   VCH
4. STOCKTON. St John. Restoration 1879   VCH
5. HUISH. St Nicholas. Restoration 1879   WRO

NEW CHURCH AT EAST GRAFTON, WILTS.

East Grafton Church from the Antiquarian and Architectural Year Book 1844,
H. Dudley printer (WCC Photographs and Prints collection)

FERREY, Edmund architect
1. SWINDON. Edgeware Road, St Paul's, 1881 (demolished 1965) P (But JB says
   by Sir A. Blomfield)

FIELD & HILTON architects of London
1. WINTERBOURNE BASSETT. St Katherine's Church, chancel, porch & roof
   JO   East window 1857   WRO

- 39 -

FIGES, William & Co ironfounders of Salisbury
Ironmongers at Blue Boar Row in 1842 ETD
1. SALISBURY. Laverstock Bridge 1841

FILDES, Geoffrey architect
1. AMESBURY. Salisbury Road, Antrobus Hall 1925 (Neo-Wren style) VCH P

FILER, Martin architect
1. COLLINGBOURNE KINGSTON. Parsonage 1812   WRO

FILTON, William builder of Great Somerford
1. MALMESBURY. Workhouse 1838 (architect 'an amateur')   RG

FINDEN, John architect of Bath and London
1811 exhibited plans of alterations to Lower Assembly Rooms and West of England Club, Bath at RA. Designed Compton Castle at Compton Pauncefoot, Somerset. Fond of castellations.
1. MELKSHAM WITHOUT. 10 houses and associated buildings at Melksham Spa in Spa Road c1815   II

FISHER, Frederick Richard architect, surveyor, carpenter and builder of High Street, Salisbury   ETD 1842. Probably son of Money Fisher qv.
1. SALISBURY. Halle of John Halle, renovation 1834 (with A.W.Pugin) RCHM
2. BARFORD ST MARTIN. Parsonage 1839   WRO
3. SALISBURY. Fisherton Anger. School site plan 1844   WRO
4. ALLINGTON. St John the Baptist 1848-51   WRO
5. SALISBURY. High Street, George Inn, drew plan of medieval building c1850 RCHM
6. COOMBE BISSETT. Parsonage 1852   WRO
7. MERE. Parsonage 1863   WRO
8. SHREWTON. Parsonage 1870   WRO

FISHER, Money architect, surveyor, carpenter and joiner of Salisbury
At High Street in 1830. ETD Also **Money FISHER & SON** in 1830. ETD
1. WILTON. Wilton House. Completed James Wyatt's alterations after Wyatt's dismissal in 1810   RCHM
2. TROWBRIDGE. Parsonage 1812   WRO
3. CHILMARK. Parsonage 1814   WRO

FLEMING, builder
1. AMESBURY. Workhouse 1837 (architects Scott and Moffatt)   RG

FLETCHER, David architect of Bristol
1. SALISBURY. St Thomas. Restoration work 1938-9   WRO

FLITCROFT, Henry architect 1697-1769
A protege of Lord Burlington, hence his nickname Burlington Harry. Succeeded as Master Mason and Deputy Surveyor in the Office of Works. Second generation neo-Palladian. RCHM "Competent but uninspired".
F,H & P
1. GREAT BEDWYN. Tottenham Park. Carried out designs of Boyle (Lord Burlington) c1720 onwards   RCHM
2. AMESBURY. Amesbury Abbey. Survey of estate 1726. Addition of wings to left and right of facade c1730s. Possibly alterations to Kent House by

1761. Possibly Gay's Cave (a grotto)   RCHM
3. STOURTON WITH GASPER. Stourhead. Temple of Ceres (later Flora) in the
   grounds 1744-5
4. STOURTON WITH GASPER. Stourhead. Pantheon 1754-6   RCHM
5. STOURTON with GASPER. Stourhead. Temple of the Sun 1765   RCHM
6. STOURTON WITH GASPER. Stourhead. Alfred's Tower c1765   RCHM

FLOOKS, John Harris architect and surveyor of Wilton
Building and land surveyor at Burdensball in 1830. ETD
1. AMESBURY. Parsonage 1824   WRO
2. AMESBURY. Amesbury Abbey. Plans of house with proposed alterations 1831
   but not employed (RIBA Drawings Collection)   RCHM

FODEN, S.P architect of 31 Essex Street, Strand, London
1. STRATTON ST MARGARET. Highworth Road, Highworth and Swindon Union
   workhouse (St Margaret's Hospital) 1846   RG

FOGG, Thomas H architect
1. CHIPPENHAM WITHOUT. Sheldon Manor, alts. 1910-11   JO

FOLEY, John Howard architect
1. BROUGHTON GIFFORD. School, addition 1872   WRO

FOLEY, SON & MUNDY auctioneers, surveyors and land agents of Trowbridge
Established 1845. H.Foley was Town Surveyor of Bradford-on-Avon in 1887. YP
(See also MUNDY.)

FORD, Henry architect
1. WILTON. Fugglestone. St Peter. Parsonage 1812   WRO

FORSYTH, W.A architect
1. CHILTON FOLIAT. St Mary. Restoration work in 1919, 1932 & 1933
   WRO

FORSYTH & MAULE architects
1. CHILTON FOLIAT. St Mary. Restoration work 1924, 1928   WRO

FORT, Alexander joiner to the Office of Works. Died 1706.
Master mason employed by Wren.   DOE
1. PITTON AND FARLEY. Farley, All Saints 1689/90 (probable ascription) P
2. PITTON AND FARLEY. Farley, Almshouses   C

FOSTER, architect of Harley Street, London
1. LIMPLEY STOKE. 17 all to 20 Middle Stoke called Fosters
   Buildings. Foster's family bought and rebuilt in 1840.
   Foster may have designed block.   ifo

FOSTER, John architect of Bristol
1. HOLT. Parsonage 1877   WRO

FOSTER, W architect of Tetbury, Glos.
1. LEA AND CLEVERTON. Garsdon Vicarage. Altered or rebuilt 1815
   WRO   JO

- 41 -

FOSTER & WOOD architects of Bristol
1. CORSHAM. School 1871   WRO

FOWLER, Charles architect 1791-1867
Apprenticed in Exeter, came to London in 1814 to David Laing's office.
Practiced on his own c1818 onwards. Best known for Covent Garden Market.
C
1. TEFFONT. Teffont Evias. St Michael & All Angels 1824-6   P

FOWLER, J architect of Louth, (Lincs?) 1791-1882
1. ODSTOCK. Vicarage 1869   JO
2. ODSTOCK. St Mary. A rebuilding 1870   P

FREESTONE, Roger carpenter of Westbury Leigh, Westbury
1. DILTON MARSH (or WESTBURY). Old Dilton. Addition to house c1531 RC

FRITH, C.S architect of Salisbury
1. BISHOPSTONE S. St Mary. Restoration work 1938   WRO

FRYER, carpenter
1. SEMINGTON. Whaddon House. Work 1668   LP

FULLER, Thomas architect of Bath
Emigrated shortly after designing Bradford Town Hall and won competition
for Houses of Parliament, Ottawa and became Chief Architect of Canada.
JO
1. BRADFORD-ON-AVON. Town Hall 1855   P
2. BRADFORD-ON-AVON. Cemetery 1857. 2 chapels and Gothic lodge JO

The Cemetery

FUTCHER, Robert builder of Salisbury
1. SALISBURY. Devizes Road, Police Station and lock-up 1859. Contract
   WRO

GABRIEL, Charles H architect of Calne   M
1. FYFIELD. St Nicholas. Restoration 1849   VCH
2. KINGTON LANGLEY. St Peter 1855-6 (builder Miller)   P
3. CHERHILL. Yatesbury Church. Restored 1855   JO

GABRIEL, S.B architect of Bristol
1. LACOCK. Bowden Hill, St Anne 1856   P
2. NETTLETON. West Kington. St Mary 1856   P
3. BEECHINGSTOKE. St Stephen. Additions and restorations,
   1860-1   P
4. MANNINGFORD. Manningford Abbotts church. Rebuilding 1861-4 P
5. CHERHILL. St James, restoration work 1863   WRO

GAIGER, John builder of Devizes
London born, apprenticed as bricklayer to Harry Chivers of W. E Chivers & Sons. 1946 started own business in Devizes. Joined by 3 brothers. Later acquired Northgate Street premises. Died 1984. (Wilts Gazette article).

GALE, John and BANKS, George builders of Chippenham and Lacock
1. LACOCK. Workhouse 1833   RG

GALPIN, architect of Oxford
1. CRICKLADE. High Street, St Mary. Restoration 1862-3   P
2. CRICKLADE. St Sampson's Church. Restoration 1863   JO

GAMBIER PARRY, S architect
1. ALDERBURY. St Mary. Restoration work 1912   WRO
2. ODSTOCK. Longford Castle. Bridge. Completed 1914 P

GAMBLE & WHICHCORD architects
1. DEVIZES. A theatre on the Island on the Green c1792 (demolished 1957)
   VCH

GAMES, T architect?
1. WARMINSTER. East Street, Yard House. Extensive alterations 1786.
   Drawings HA

GANE BROTHERS (C & P) architects of Trowbridge
1. TROWBRIDGE. Polebarn Road, Woollen factory 1860   P
2. TROWBRIDGE. Stallard Street, Studley Mill 1860   JO
3. TROWBRIDGE. Court Street, Mill 1862   P
4. COULSTON. St Thomas Beckett. Restoration work 1868 WRO

GANE, Richard architect   -c1877
From a family of Trowbridge mill builders. Trained as an architect and joined C.E.Giles of Taunton c1870. C1877 emigrated to Australia and died on arrival at Sydney.   JO
1. EAST KENNETT. Christ Church 1864   P
2. TROWBRIDGE. Holy Trinity Infants School c1870-5   JO
3. BRADFORD-ON-AVON. Abbey Mill 1875 (built as a cloth factory; restored

and converted by Thurlow, Lucas & Janes 1968-72). P "One of the finest mill buildings of its time ...Gothic detail applied without ostentation" JO

GANE, Charles and GANE, Richard surveyors of Trowbridge
Charles Gane, carpenter, joiner and builder of Back Street in 1822-3. ETD Ganes had yard in Church Street (later W. Smith's) on what became Tabernacle site. ML
1. SEMINGTON. Melksham Union Workhouse 1838 (architect Kendall) RG WRO

GATER, Caleb William surveyor, 1852 - ?
Lived at The Grange, Winterbourne Dauntsey. Came to Salisbury in 1873. Biography and photo. in D

GAYE, H architect
1. CHIPPENHAM. St Paul's Church. Choir vestry and organ chamber extension
     1911   SPC

GEORGE, canon, FRIBA
1. DEVIZES. St Peter, restoration work 1935-7   WRO

GIBBERD, Sir Frederick architect 1908 -
One of the first in England to accept the International Style of the 1930s. Best known for his buildings in the new town of Harlow, 1947 onwards. F,H&P Swindon Corporation consultant architect.
1. SWINDON. Fleming Way. Shopping centre (by Shingler Risdon; Gibberd
     consultant archict). Plans accepted 1960   P

GIBBS, Alexander artist   1832-86
Son of I.A.Gibbs (qv) but better known than his father. Worked for William Butterfield.   JO
1. HEYTESBURY. Church glass. Some of his best.   JO
2. GRITTLETON. Littleton Drew, All Saints Church, E window 1856   JO
3. WINTERBOURNE MONKTON. St Mary Magdalene Church, E window   JO

GIBBS, I.A artist 1802-1851
1. CASTLE COMBE. St Andrews Church. S chancel - attributed to A. Gibbs by
     Kelly 1867, but G. P. Scrope 'History of Castle Combe' (1851) says that
     the window is by the late Mr Gibbs, so presumably I. A. Gibbs. JO

GIBSON, John architect 1817-1892
Architect to N.P.B. from 1864.
1. SALISBURY. National Provincial Bank   JO

GILBERT, Sir Alfred sculptor   1854-1934
Born in London, studied in France and Italy. Work of remarkable simplicity and grace. CH Best known for Eros, Piccadilly and Clarence memorial, St. Georges Chapel, Windsor.
1. LONGBRIDGE DEVERILL. Church font c1890. Art Nouveau   JO

GILES & GANE architects of London
1. SEMINGTON. St George. Work done 1876   WRO

GILES & ROBINSON architects
Is this 'C.E.Giles, prolific West Country church architect in 1860s, earlier in partnership with Richard Carver'? NJ
1. STEEPLE ASHTON. St Mary. Restoration work 1868   WRO

GILL, J.E of Bath
1. BRADFORD-ON-AVON. Holy Trinity Church restoration 1862-6 GL JO

GIMSON, Humphrey M architect 1890-1982
Educated Bedales and UCL. Asst. to Lutyens 1911-14. With Sedding & Stallybrass, Plymouth 1920-22, and Norman Jewson, Cirencester, 1923-4. Own practice at 1 St John Street, Devizes 1925-50. Committee member of SPAB.
1. DEVIZES. St John. Restoration work 1932   WRO
2. PEWSEY. St John. Restoration work 1934   WRO
3. ALDBOURNE. St Michael. Restoration work 1934   WRO
4. CHIRTON. St John. Restoration work 1936   WRO
5. BROMHAM. St Nicholas. Restoration work 1937   WRO
6. WHITEPARISH. Restoration of cottages 1937
7. ALTON. Alton Barnes, St Mary. Restoration work 1938   WRO
8. URCHFONT. Restoration of cottages 1938
9. LITTLE CHEVERELL. East Sands 1939
10. BISHOPS CANNINGS. Pumping Station 1945 (now demolished)
11. DEVIZES. Bath Road, Maryes Cottage
12. DEVIZES. Hospital. A pair of cottages
13. DEVIZES. The Fairway, Seven Gables
14. WILCOT. Oare. Parsonage House, studio in garden

GLASCODINE, Joseph architect of Bristol
1. WARMINSTER. St. Boniface College. Glascodine built the original house of 1796, which became the core of the College, founded 1860. P

GOATER, Charles H architect
Practice at 6a Silver Street, Trowbridge (Kelly's 1923) over the then Wiltshire Times office.

GODFREY, C architect
1. SEAGRY. Lower Seagry. Church Farm, alterations and extensions 1947 VCH

GODWIN, Edward William architect 1833-1886
Practised in Bristol and London.   DNB
1. BOX. Ditteridge. St Christopher. Restoration 1860   P

GODWIN, F architect of Lyneham
1. TOCKENHAM. Vicarage 1897   WRO   JO

GOOCH, Sir Daniel engineer of Swindon 1816-1889
Born at Bedlington, Northumberland. GWR locomotive superintendant 1837-64. Gooch may have designed various GWR buildings at Swindon along with Brunel (qv).
1. SWINDON. Rodbourne Road, GWR Works Manager's Office 1841 "Presumably designed, built and used by Gooch" DOE

GOODERICK, Matthew painter died c1654
1. WILTON. Wilton House. Ceiling painting. Mid 17thC   RCHM

- 45 -

GOODRIDGE, Alfred S architect 1828-1915
Son of Henry Goodridge. Number of commissions in Trowbridge through Sir
Roger Brown, clothier and municipal benefactor.   JO
1. TROWBRIDGE. Hilperton Road, Highfield 1859 P (KR says by Manners and
   Gill)
2. TROWBRIDGE. Town Hall 1887-9   P
3. TROWBRIDGE. Polebarn Road, Lady Brown's Almshouses 1900
4. TROWBRIDGE. Cemetery, Brown mausoleum (?)   JO

GOODRIDGE, Henry Edmund builder and architect of Bath 1797-1863
Son of James Goodridge, local builder and agent to Darlington estate until
1835. Apprenticed to John Lowder c1817. NJ   Greek Revival in Bath, but
Gothic churches also. Well known for Beckford's Lansdown Tower, Bath. Most
significant Bath-based architect of 19thC. Surveyor to Great Western
Railway. NJ
1. SOUTH WRAXALL. St James, galleries 1823   C
2. MALMESBURY. Abbey. Replacement window 1823, W window c1830 VCH P
8. LACOCK. Notton House. Refront 1830-40 (possibly H.E.G)   JO
3. ROWDE. St Matthew. Rebuilding, except for tower 1831-3   P
   Drawings in the RIBA Drawings Collection given by Mowbray Green who
   said that they had been drawn by H.L. Elmes, later architect of St.
   George's Hall, Liverpool, while a pupil.
4. ATWORTH. St Michael. Rebuilding 1832 (again rebuilt in 1852, but
   not by Goodridge)   PC
5. DEVIZES. Castle. Round tower, after 1838, the beginning of the
   Victorian building of the castle after the site was acquired by the
   Leach family in 1838.   P
6. NORTH BRADLEY. Christ Church   VCH
7. ATWORTH. Cottles (now Stonar School). Extension   JO
9. COLERNE. Rectory 1842 (Italianate style of H.E.G?)   VH2
10. CORSHAM. Pickwick. Bath Road, St. Patricks R C Church 1848 JO
11. CORSHAM. Pickwick. School 1857   WRO

GOOLD, Vivian architect
1. KINGTON LANGLEY. Old Chapel Field for Robin and Heather Tanner 1931
   JO

GORDON, George H architect of London
1. WILTON. Parsonage 1882   WRO

GOULD, stone mason of Swindon
Produced handsome tablets in country churches c1810. He or his son may have
designed simple classic stone houses at Swindon. JB

GOULD, John architect of Tottenham Park, Great Bedwyn
1. MARLBOROUGH. National School 1852-3 (Builder 1852 p203)   JO
2. LITTLE BEDWYN. School 1853   WRO

GOVER, William architect
1. HINDON. St John's Church. Enlarging 1836 (but rebuilt 1870-1 by
   T.H.Wyatt)   VCH

GRANT, William mason of Bradford-on-Avon
Leasing cottage at Newtown quarry in 1657   MP

GREEN, LLOYD & SON architects (later **GREEN, LLOYD & ADAMS**)
Practice at Bedford Row, London.
1. SALISBURY. Close. Training College. post-war extensions 1949
   onwards   P
2. CALNE. St Mary's Schools. Additions, viz. science block 1961, boarding
   house 1963, chapel 1971   P
3. SALIBURY. Close. Audley House, extension to kitchen and dining
   room 1965   P
4. SALISBURY. Old Deanery. Restoration   RIBA

GREENMAN, George carpenter
1. DAUNTSEY. Rectory 1829-33   DR

GREENMAN, William mason
1. STANTON ST QUINTIN. Upper Stanton Farm, repairs 1836   EG

GREENSHIELDS, Thomas architect of Oxford
1. MARKET LAVINGTON. Parsonage 1841   WRO

GRIFFITH, Edward H architect
Practice in Downton (Kelly's 1927)

GUNSTON, Samuel mason
1. DAUNTSEY. Rectory 1829-33   DR

GUNSTON, William mason
1. DAUNTSEY. Rectory 1829-33   DR

GUNSTONE, John architect or builder
1. MELKSHAM. National School, Infants 1870   WRO

GYE, thatcher
1. DAUNTSEY. Rectory, barn re-thatched c1833   DR

HABERSHON, W.G and PITE architects of Bloomsbury Square, London
1. TROWBRIDGE. St Thomas's Infants School 1872   WRO

HADEN & CO heating engineers of Trowbridge
George Haden came to Trowbridge in 1814 working for Boulton and Watt.
Settled in town as mill engineer 1815. Joined by brother James. Firm became
world-famous for steam and warm air heating systems.   KR
1. CHIPPENHAM. St Paul's Church. Heating apparatus 1915   SPC

HAKEWILL, John Henry architect of 50 Maddox Street, London 1811-1880
Son of Henry Hakewill, architect 1771-1830   DNB
1. URCHFONT. St Michael. Restoration work 1839   WRO
2. URCHFONT. Vicarage altered or rebuilt 1839   JO
3. STERT. St James Church. Rebuilding 1845-6   P
4. SEAGRY. St Mary's Church, rebuilding 1849   VCH   P
5. STANTON ST QUINTIN. St Giles Church. Upper part of tower (probably
   Hakewill 1851)   P
6. SUTTON BENGER. All Saints Church. Restoration 1851 "Cruelly
   restored" - P   (1849 VCH)
7. KINGTON ST MICHAEL. St Michael & All Angels. Restoration 1858

- 47 -

"Terribly over-restored" - P
8. CORSHAM. Corsham Side National School and house 1864   WRO
9. GREAT SOMERFORD. SS Peter & Paul. Restoration 1865   VCH P
10.CORSHAM. Neston. SS Philip & James 1866   P

HALE & Son builders of Salisbury
Frances Hale, Raymon Henry Hale and Edward Hale   RG
1. ALDERBURY. Workhouse 1879 (architects Nichols then Hall) RG

HALL, Henry architect of 15 Duke Street, Adelphi, London 1826-1909 RG
1. SALISBURY. Queen Street. Pinkney's Bank 1879   RCHM
   (presumably Pevsner's Royal Insurance Building 1878)   P
2. ALDERBURY. Workhouse 1879 (succeeding Nichols)   RG

HALL, L. K architect
1. MELKSHAM. St Andrew's Church. Restoration work 1934   WRO

HALL, Roger carpenter   still alive 1699
1. DILTON MARSH (or WESTBURY). Old Dilton. Extension to Church House late
   17thC RC

HALLIDAY & ROGER architects of Cardiff
1. SWINDON. Swindon Hospital   JO

HALLIDAY, John Edmund clothier, dyer and architect of Warminster
Owned much property in Warminster and elsewhere. Family long established in
Warminster and Taunton, Somerset. A "competent architect". Notebook at WRO
has fine architectural drawings. Designed new parsonage house at Coleford,
Som. 1832. Resigned from Warminster Highways Board 1843 as health poor. His
father Edmund, a wine merchant, may also have been an amateur architect.
1. WARMINSTER. East Street, Yard House. Alterations 1833 after he inherited
   the house   HA
2. WARMINSTER. Cold Harbour. Cottages rebuilt 1835   HA

HAMMOND, John architect
1. MARLBOROUGH. Town Hall. Rebuilding 1792-3 (again rebuilt by Ponting
   in 1901-2)   VCH

HANNAM, Joseph surveyor
1. DAUNTSEY. Rectory, measuring up masons' work 1829-33   DR

HANNEY, John the elder tiler of Bradford-on-Avon
Leasing cottage at The Grove above Newtown in 1632.   MP

HANSOM, Charles Francis architect   1816-
Leading Roman Catholic architect. His brother Joseph was at one time in
partnership with Pugin. NJ
1. DEVIZES. Church of the Immaculate Conception 1865   P
2. LYNEHAM. Bradenstoke. St Mary's Church 1866   P
3. CORSHAM. St Bartholomew's Church. Restoration 1874/8 with the addition
   of the Methuen Chapel   P

HANSOM and WELCH architects of Birmingham
1. WOODFORD. Parsonage. Early 19thC   WRO

HARDICK, E.E
1. DEVIZES. Sheep Street, Baptist Chapel 1851-2   P

HARDICK, Thomas carpenter and joiner of Warminster
At Townsend in 1830 ETD and Boreham Road as carpenter and builder in 1842 ETD
1. WARMINSTER. East Street. Turnpike house 1828. Estimate for removal of same building to Boreham Road by Thos. Hardick & Son not accepted in 1840   HA

HARDICK, W architect or surveyor of Warminster
In the High Street as timber surveyor in 1842 ETD
1. WYLYE. Fisherton Delamere. St Nicholas Church. Rebuilding of chancel 1862   P

HARDICK, William Henry architect
Practice at 25 High Street, Warminster (Kelly's 1889-1907)
1. NORTON BAVANT. All Saints. Parsonage 1838   WRO

HARDICK, W & Son, surveyors of Warminster
1. WYLYE. Fisherton de la Mere, National School and house 1872   WRO
2. BOYTON. National School 1874   plan, WRO

HARDING, John & Son architects of 58 High Street, Salisbury (Brown's Directory 1913 and see J. Harding below.)
1. SALISBURY. St Martin's Church. Restoration work 1911   WRO
2. DONHEAD ST MARY. St Mary' Church. Restoration work 1911   WRO
3. SALISBURY. St Thomas's Church. Restoration work 1912   WRO
4. ORCHESTON. St Mary's Church. Restoration work 1915   WRO
5. ANSTY. St James' Church. Restoration work 1917   WRO
6. SALISBURY. East Harnham. St George's Church. Restoration work 1926 WRO

HARDING, John architect and surveyor
Practice at 51 Canal, Salisbury (Kelly's 1875-1899); 58 High Street, Salisbury (Kelly's 1911-1923 at least)
1. BISHOPSTONE S. School house for National School 1854   WRO
2. SALISBURY. 40 St Ann Street (former Salisbury Museum). Brick facade 1863, on C18 house.   P
3. SHREWTON. National School and house 1868   WRO
4. ODSTOCK. St Mary's Church. Restoration work, presumably before the 1870 rebuilding.   WRO
5. WINTERSLOW. All Saints Church. Restoration work 1871   WRO
6. SALISBURY. Off Milford Street, Ragged School 1873   WRO
7. ALLINGTON. St John's Church. Restoration work 1876   WRO
8. ALLINGTON. Rectory, rebuilt c1880   WRO
9. TEFFONT. Teffont Magna, National School 1877   WRO
10. DILTON MARSH. Holy Trinity Church. Restoration work 1880   WRO
11. WESTBURY. Vicarage 1880   JO
12. ODSTOCK. Vicarage c1880   JO
13. STOCKTON. Vicarage altered or rebuilt c1880   JO
14. SALISBURY. St Martin's Church. Restoration work 1884   WRO
15. WESTBURY. All Saints Church. Restoration work 1884   WRO
16. STOCKTON. St John's Church. Restoration work 1885   WRO
17. WOODFORD. All Saint's Church. Restoration work 1894   WRO

18. FONTHILL BISHOP. Restoration work 1903-12   WRO
19. SALISBURY. St Thomas's Church. Restoration work 1922   WRO
20. SALISBURY. Winchester Street, house and shop   JO

PROPOSED—RECTORY—HOUSE

—ALLINGTON—WILTS—

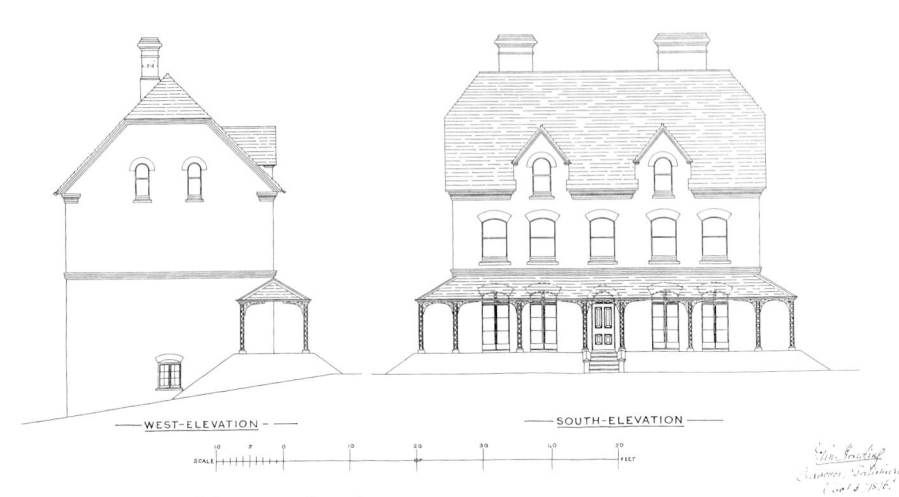

—WEST-ELEVATION—   —SOUTH-ELEVATION—

HARDING, M architect of Salisbury
1. WHITEPARISH. Parsonage 1879. Altered or rebuilt   WRO, JO
   (May be his father's)
2. NORTH TIDWORTH. Holy Trinity Church. Restoration work 1912   WRO
3. NORTON BAVANT. Parsonage 1912-1921   WRO
4. PITTON AND FARLEY. Farley. All Saints Church. Restoration work 1927
   WRO
5. ODSTOCK. St Mary's Church. Restoration work 1927   WRO
6. FOVANT. St George's Church. Restoration work 1931   WRO
7. EBBESBOURNE WAKE. Fyfield Bavant. St Martin's Church. Restoration work
   1931   WRO
8. ODSTOCK. Nunton. St Andrew's Church. Restoration work 1935   WRO
9. CHAPMANSLADE. SS Philip & James. Restoration work
   1935-6   WRO
10. FIGHELDEAN. St Michael's Church. Restoration work 1936   WRO
11. BERWICK ST JAMES. St John's Church. Restoration work 1936   WRO

HARDING AND ELGAR architects of Salisbury
1. BURCOMBE WITHOUT. St John the Baptist Church. Restoration work 1937
   WRO
2. BRIXTON DEVERILL. St Michael's Church. Restoration work 1937 WRO
3. LONGBRIDGE DEVERILL. SS Peter and Paul Church. Restoration work
   1938   WRO
4. REDLYNCH. St Birinus Church. Restoration work 1938   WRO
5. BROADCHALKE. All Saints Church. Restoration work 1939   WRO

HARDMAN, J stained glass maker
1. SALISBURY. St Osmund R.C. E and S windows, designed by Pugin 1850 P

- 50 -

2. BRADFORD-ON-AVON. Holy Trinity Church. W window c1862   JO
3. BREMHILL. St Martin. W window 1864   P
4. COMPTON BASSETT. St Swithin. E window c1866   P
5. GREAT BEDWYN. St Mary. S transept glass, designed by Street   P
6. WOOTTON BASSETT. St Bartholomew and All Saints. E window   P

HARDWICK, Philip architect 1792-1870
Stopped practising in the 1840s. Best known for his Euston Station. F, H & P
1. CORSHAM. Hartham Park. Chapel in the grounds built in 1862 to Hardwick's earlier plans.   P

HARDY & SON
1. CHUTE. National School, drawing 1858 (architect Wm White)   WRO

HARRIS, John builder
1. CHISELDON. National School and house 1870   WRO

HARRIS, Thomas plumber, painter and glazier of George Street, Warminster
1. WARMINSTER. Workhouse fittings 1836-7   RG

HARRISON, H.B architect
1. SEAGRY. Seagry House. Brick stables 1914   VCH

HARRISON, Henry architect c1785-1865
Is this the Henry Harrison who practised from Park Street, Grosvenor Square?   C
1. STEEPLE ASHTON. Old Parsonage alterations 1829   P
2. MARLBOROUGH. Rectory 1832-3 (rebuild)   VCH C p393 JO

HART, Mungo architect
1. CALNE WITHOUT. Derry Hill, National School and house 1873   WRO

HARVEY, stone carver
1. SEMINGTON. Whaddon House. Work 1684-5 and may be the carver of Bath paid for lions' heads (probably for gateway) in 1682   LP

HARVEY, Edwin builder of Swindon 1843-1925
Born Keynsham. At Park Hotel, Swindon 1878-81, at 71 William Street 1883-90 and at 50 Exmouth Street 1891-1925 (with yard at rear).   TS
1. SWINDON. Park Tavern, William Street 1878   TS
2. SWINDON. Terrace of shops, west side of Cambria Bridge Road 1878 TS
3. SWINDON. 160-170 Clifton Street 1878   TS
4. SWINDON. 170-180 Clifton Street, early 1880s   TS
5. SWINDON. 8-21 Exmouth Street, early 1880s   TS
6. SWINDON. 56-64 Exmouth Street, early 1880s   TS
7. SWINDON. 1-19 William Street, early 1880s   TS
Entries 4-7, architect was William Drew q.v.

HARVEY, F.W architect and builder
1. CHILMARK. Addition to National School 1859   WRO

HAYTER, James quarry master and mason of Box
Harrods Directory 1865.

HAYWARD, John carpenter
1. DAUNTSEY. Rectory 1829-33   DR

HEMMING, Alfred O stained glass designer
1. CHIPPENHAM. St Paul's Church. E chancel window 1905   SPC

HENLEY, G.R builder of Swindon
Firm established 1856, advertising 1882 as at 4 Cricklade Street, King William Street and Prospect Place in Deacon's Gazetteer.

HENLY, Henry C architect
Practice at High Street, Calne (Kelly's 1880); Curzon Street, Calne (Kelly's 1889-1899)

HENSHAW, F architect of Andover, Hants.
1. LUDGERSHALL. St James Church. Restoration work 1936   WRO

HEPWORTH, P.D architect
1. TROWBRIDGE. County Hall 1938-40 (builders J. Long & Sons of Bath). The first plan had no attic storey.   KR VCH

HERBERT, Lord Henry architect of Wilton House, Wilton
9th Earl of Pembroke. Like Boyle called the 'Architect Earl'.
1. WILTON. Wilton House. Palladian Bridge. Joint designer with Roger Morris 1736   RCHM

HICKS, J architect of Bristol
1. MONKTON FARLEIGH. Rectory 1844-6   P

HICKS, James architect of Redruth, Cornwall
1. BOX. Fogleigh House for C. J. Pictor. (The Architect 9.7.1881) Stone Gothic house built for quarry owner, similar to Rudloe Park Hotel & ?Gastard House, near Corsham.   JO
2. BOX. London Road, Clift House. Clift Quarry Offices, Gothic c1870, belonged to Pictor family.   JO
3. BOX. Village Schools 1874-5. (Bath Chronicle 16.11.1875)   JO
4. BOX. Rudloe Park Hotel c1875, for Pictors   JO

HILL, H.L.G architect
1. BOYTON. St Mary's Church. Restoration work 1937   WRO

HILL, W architect of Leeds
1. DEVIZES. Corn Exchange 1857 P (Competition held: C.J.Phipps 1st prize; W.Hill 2nd)   JO

HIND, Richard mason
1. DAUNTSEY. Rectory 1829-33   DR

HINE, G.T & PEGG, H Carter architects
Specialist asylum architects. Hine became consulting architect to Commissioners in Lunacy for England. In 1889 he had used compact echelon

- 52 -

plan for Claybury Hospital, London which was much imitated. GS RCHM
1. ROUNDWAY. County Asylum. The Annexe c1913 (builder Chivers) RCHM

HITCHCOCK, William mason
1. LACOCK. Lacock Abbey, alterations 1754-5 (architect Miller)   PC

HOARE, bricklayer
1. FONTHILL GIFFORD. Fonthill Abbey. The Palladian house built after the
   fire of 1755   P

HOLDOWAY, T. & Sons Ltd builders of Upper Eden Vale, Westbury
1. BRATTON. Court House, repairs 1947 (architect Vallis)   WRO

HOLLAND, Henry builder
Father of the better-known architect of the same name (1746-1806).
1. CALNE WITHOUT. Bowood House. Carried out designs of Adams brothers
   c1761-3   WAM 41

HOLLOWAY, H Thomas architect and surveyor
Practice at 36 Marshfield Road, Chippenham (Kelly's 1907); as **HOLLOWAY &
FOGG** at 7 Market Place (Kelly's 1911, 1915). Fogg continued after
Holloway's death but was killed in 1914-18 war. W. Rudman (qv) acquired the
practice in c1921. Holloway was responsible for designing many of the older
shops in Chippenham High Street and apparently also had a great deal to do
with the original sewerage scheme in Chippenham. Large scale OS Survey
Sheets on which these sewers were plotted inherited by Edwards & Webster.
PE

HOOKE, Robert architect
1. RAMSBURY. Manor 1681 - "almost certainly the architect"   VCH

HOOPER & DOBBIN architects of Westminster, London
1. WOODFORD. School and house 1873   WRO

HOPPER, Thomas architect 1776-1856 of Sovereign Street, Connaught Terrace,
Edgeware Road, London in 1828.   PL   Born Rochester, where his father
was a surveyor. The Prince Regent was his patron, and employed him to make
alterations to Carlton House. The royal patronage brought him much country
house work. "Thoroughgoing eclectic". "Successful and fashionable
architect." RCHM C
1. SALISBURY. Guildhall. Alterations 1828-30 - N portico rebuilt
   in projecting form with a room above it for the grand jury. Alterations
   to basement, railings outside   PL   RCHM
2. AMESBURY. Amesbury Abbey. Grandiose rebuilding for Sir E.Antrobus Bt
   from 1834-41. Design exhibited R. A. 1841 (cf also Country Life
   1.3.1902). Used steel beams and plate glass   RCHM   JO
3. WEST ASHTON. Rood Ashton. Altered for the Long family 1836
   (originally Wyatville 1808). Now a reconstructed ruin   P
4. WEST ASHTON. Rood Ashton. Rood Ashton Lodge -lodges with carriage house
   and stables c1830 for Long Family. "Possibly by Hopper" DOE

HORWOOD Brothers glaziers of Frome, Somerset
1. TROWBRIDGE. Holy Trinity Church, glass of E Chancel window 1861 JO
2. CHAPMANSLADE. SS Philip and James Church (architect Street)   JO

HUGALL, J. W architect of Pontefract, Yorks.
1. DURRINGTON. All Saints Church 1851   P
2. FIGHELDEAN. St Michael's Church. Tower top ("rather distressingly neo-
    Norman" -P) 1851, restoration 1859-60   VCH
3. STANTON FITZWARREN. St Leonard's Church. Restoration 1865   P
4. HIGHWORTH. St Michael's Church   WRO

HUGHES, R.N architect of 6 Winckley Street, Preston
1. TISBURY. Wardour School 1859   (Arundell collection)   WRO

HULBERT, Walter Frederick Edmund builder of Bradford-on-Avon
1. BRADFORD-ON-AVON. 9 Belcombe Place. Empty plot bought by Hulbert in
    1906 and house built c.1907, probably for himself.   PMS

HUMBY, William builder of Wilton
1. WILTON. Workhouse 1837 (architect Hunt)   RG

HUNT, Edward surveyor of New Alresford, Hants.
1. ALDERBURY. Workhouse 1837 (pulled down 1879)   RG
2. WILTON. Workhouse 1837   RG

HUNT, F.W architect
1. CORSLEY. St Margaret's Church (1833 John Leachman). Alterations by
    Hunt 1891   P

HUNT, H.R carpenter
1. CHIPPENHAM. St Paul's Church, woodwork in children's corner 1949, and
    baptistry 1955 (architect O. Brakspear). Choir stalls c1949-51   SPC

HURLE builder
1. BRATTON. National School 1846   WRO

HUTCHINS, Thomas surveyor
1. BISHOPSTONE S. National School 1870   WRO

IMRIE, G. Blair architect
Later IMRIE, PORTER & WAKEFIELD of Market Place, Warminster.
1. CODFORD. St Mary's. Restoration work 1934   WRO
2. CODFORD. St Peter's. Restoration work 1934   WRO

INGLELOW, Benjamin architect
1. COLLINGBOURNE DUCIS. Rectory 1863   VCH

INGLEMAN, Richard architect of Southwell
1. DEVIZES. Gaol 1810 First called the "New Bridewell", renamed in 1836
    the "New Prison" - designed on panopticon principles. Demolished   VCH
2. CHIRTON. Conock Manor. The single-storey wings and dairy c1817.   P

IRVINE, James Thomas architect
Scholarly Clerk of the Works for Sir G.G.Scott for the restorations at Bath
(1863-73) and Wells.   BC JO
1. BRADFORD-ON-AVON. St Lawrence Church. First supervisor of the
    restoration 1870s. P   Remarkable study of the Saxon chapel.   WR

- 54 -

ISBORN, Ernest Charles architect 1856-?
Southbroom Cottage, Devizes. Practice at 33 St John Street, Devizes.
Father was Chief Engineer of LSE Railway. Worked with Joseph Clarke,
London, and Sebastian Waterhouse, Liverpool. Came to Devizes 1881 (Kelly's
1907, 1911, 1915). D
1. DEVIZES. Bath Road, factory for Central Wilts Bacon Co. Ltd. D

JACOB, H architect of Salisbury
1. LAVERSTOCK. St Andrew's Church. Restoration work 1939 WRO

JACOBSEN, Theodore merchant and amateur architect Died 1772
1. ODSTOCK. Longford Castle. Loggia c1757 (either Jacobsen's design, or
   that of Roger Morris) P (Perhaps advice on design, for Sir Jacob
   Bouverie.) C

JAMES, John architect and carpenter
1. WILTON. Church. Carpentry c1700-1710 RCHM
2. WILTON. Wilton House. Alterations after 1705 fire in N range "might be
   attributed to James". RCHM

JONES AND ATWOOD architects of Stourbridge, Worcs.
1. WESTWOOD. Avoncliff. Workhouse. Alterations 1898 RG

JONES AND CO. of Mount Street, Birmingham
1. CLARENDON PARK. Clarendon House, conservatory 1828 PL

JONES, Daniel builder of Bradford-on-Avon
1. CHIPPENHAM. St Paul's Church 1853 (architect G.G.Scott) SPC

JONES, Daniel & JONES, Charles builders of Tory, Bradford-on-Avon
Charles Jones became a Town Commissioner for Bradford-on-Avon in 1839. An
eminent building firm. YP Sound and skilful workmen. GL Had quarries at
Tory and Jones's Hill.
1. BRADFORD-ON-AVON. Christ Church 1841 (architect Manners) YP
2. WINSLEY. St Nicholas Church. Alterations 1841 GL
3. WILTON. St Mary and St Nicholas Church 1843 (1840 YP) (architects Wyatt
   and Brandon)
4. MELKSHAM. Town Hall 1847 JO
5. BRADFORD-ON-AVON. Bath Road, Frankleigh House. Rebuilding on old core
   (architect H. Clutton) 1848 WRO
6. BRADFORD-ON-AVON. Trowbridge Road, Victoria and Albert Villas GL

JONES, Inigo architect and stage designer 1573-1652
The bringer of the classical style to England, and by 1614, Palladianism.
Surveyor of the King's Works from 1615 until the Civil War in 1642.
1. FONTHILL BISHOP. Fonthill Abbey - gateway from Fonthill Bishop
   (attrib.) P
2. WILTON. Wilton House, S front by his assistant Isaac de Caus q.v.
   c1632 with his advice. RCHM F,H &P
   (With John Webb) design for the new house after the fire of 1647
   (E front had survived the fire, and this was incorporated into Jones'
   design). RCHM VCH

JONES, Isaac mason of Bradford-on-Avon   fl.1831-1864
1. BRADFORD-ON-AVON. Belcombe Place. Nos 15 and 16. Plot bought by John Jones (qv) by 1822 and sold to I.Jones in 1831. Houses built 1836-7.   PMS

JONES, John mason of Bradford-on-Avon
Of Wine Street in 1822. ETD   Family had extensive property and quarry interests in Wine Street, Tory and Newtown in the 19thC.   PMS
1. BRADFORD-ON-AVON. The Grove (above Newtown). 1791 P. Methuen leased close called The Grove to Jones for 99 years. "6 Messuages were afterwards erected on the ground."   WAM 32

JONES, John builder of Bradford-on-Avon
Stone mason, Newtown in 1852-3 (Slater's directory).
1. BRADFORD-ON-AVON. 7 and 8 Silver Street, reconstructed 1876   HF

JOSEPH, E.M architect
1. SALISBURY. New Street, Alexandra Rooms 1950-5   P

KAYNS, wallpaperer
1. DAUNTSEY. Rectory 1833   DR

KEENE, Henry architect 1726-1776
1746 - Surveyor to the Dean and Chapter of Westminster, later Surveyor to the Fabric of the Abbey as well. Gothic style, roughly contemporary with Strawberry Hill. Classical too - see his work in Oxford.
1. CALNE WITHOUT. Bowood. Alterations and enlargements to original (1725) house c1754-60 (employed by the 1st Earl of Shelburne until the latter's death. The 2nd Earl, in 1760, called in Robert Adam).   P
2. CORSHAM. Corsham Court. Designs for extension of the house (N front, etc) 1759 (he was replaced in 1761 by Capability Brown, who still used Keene's designs).   P

KELLOW, James stone mason of Tisbury
1. SALISBURY. General Infirmary 1768   RCHM

KEMPE, C.E stained glass maker
Lived at 17 The Close, Salisbury. Trademark - the wheatsheaf. P
1. PATNEY. St Swithin's Church c1873   P
2. SALISBURY. Bemerton. Window in N aisle 1878   P
3. LUCKINGTON. Sts Mary and Ethelbert. E window c1881   P
4. STRATFORD TONEY. Sts Mary and Lawrence, E window 1884   P
5. PITTON and FARLEY. Pitton. St Peter's Church, E window 1886   P
6. CRICKLADE. St Sampson's Church. W window 1888   P
7. HILPERTON. St Michael's Church, chancel SE window 1890   P
8. CORSHAM. St Bartholomew, E window 1892 and Methuen Chapel E window 1899   P
9. WEST LAVINGTON. All Saints. Chancel 2 N windows 1892 and 1907   P
10. BISHOPSTROW. St Aldhelm's Church. NE window c1892   P
11. SALISBURY. St Ann Street. House of Mercy, glass in chapel 1895-1904 P
12. STANTON FITZWARREN. St Leonard's Church, chancel 1896 and 1907 P
13. BLUNSDON ST ANDREW. St Andrew's Church. N window c1896   P
14. MELKSHAM. St Michael's Church, S Chapel, SE window 1897 signed JO P
15. SWINDON. St Mark's Church. S chapel E window 1897 and a N aisle window P
16. DEVIZES. St James's Church, N aisle NE window c1900   P

17. BREMHILL. St Martin's Church, S aisle W windows 1903   P
18. WYLYE. St Mary's Church. N aisle NE and NW windows 1904   P
19. BISHOPSTONE N. Little Hinton. St Swithin's Church, E window c1910 P
20. SALISBURY. 17 The Close. Various glass (Kempe's house)   P
21. SALISBURY. King's House. Chapel (of Training College), glass   P

KEMPE & TOWER stained glass makers of Salisbury
C.E.Kempe qv. and W.E.Tower (partner then successor).
1. CALNE WITHOUT. Blackland. St Peter's Church. E window c1906   P
2. LATTON. St John Baptist Church. E window 1911   P
3. LUCKINGTON. Sts Mary and Ethelbert. Various windows   P

KEMPTHORNE, Sampson architect of Hants.
In New Zealand by 1840-42. C
1. WARMINSTER. Workhouse 1837   RG

KEMP-WELCH & REYNOLDS architects
Practice at St Thomas's Square, Salisbury (Kelly's 1875)

KENDALL, Henry Edward architect 1776-1875
One of the founders of RIBA, early meetings of the committee were held at his house. C Designed various workhouses. RG
1. SEMINGTON. Melksham Union workhouse (later St George's Hospital) 1838
   RG

KENNEDY, George architect
1. LUDGERSHALL. Biddesden House, gazebo in the grounds 1932 P (Same as swimming pool changing pavilion? 1932 VCH)
2. COMPTON BASSETT. Compton House, rebuilt c1932-5   P

KINGESTON, Francis architect
1. BROAD TOWN. Parsonage 1856   WRO

KINWARD, Thomas joiner
From 1660 until his death in 1682 Master Joiner of the King's Works. RCHM
1. WILTON. Wilton House. Refitting post 1647 fire (Webb and Jones architects). Document refers to 'Mr Kennard'   RCHM

KNAPP, James builder, carpenter and wheelwright of Melksham
Brother of John Knapp (qv) and mentioned in John's will of 1863. At Bath Road in 1830 and 1842 ETD PMS

KNAPP, John surveyor, timber and land agent of Belcombe Cottage (now Belcombe Croft), Bradford-on-Avon. Previously of Trowbridge Road, Bradford-on-Avon. Lived c.1792-c.1863.
1. BRADFORD-ON-AVON. Belcombe Place. Knapp set out the road and plots in 1822 and was one of the 3 owners. Reserved the corner plot for himself and may have designed his house, now Belcombe Croft. Built by 1828 PMS

KNAPP, Joseph stone tiler
1. STANTON ST QUINTIN. Upper Stanton Farm, repairs 1836   EG

LANCASTER, architect of Steeple Ashton
1. WEST ASHTON. Piggeries and cowshed at National School 1853   WRO

LANE, Josiah of Tisbury
1. CALNE WITHOUT. Bowood, grotto   P
2. FONTHILL GIFFORD. Fonthill Abbey, grotto   P
3. TISBURY. Wardour, Old Castle, grotto 1792   P

LANSDOWN, Frederick joiner of Trowbridge
Architect pupil but became architectural joiner.   ML

LANSDOWN, Thomas Smith architect of Swindon
1. TOCKENHAM. Parsonage 1866   WRO
2. PURTON. Bentham House, additions 1867   P
3. PURTON. Wesleyan Chapel 1868   JO
4. SWINDON. Old Town National School and house 1868   WRO
5. SWINDON. Faringdon Road. Conversion of the "Barracks" (a GWR lodging house) to Wesleyan Chapel 1869 (after 1962, a railway museum) VCH

LANSDOWN & SHOPLAND architects and civil engineers
Practice at 42 Cricklade Street, Swindon (Kelly's 1875, 1880) and 18 King William Street (Kelly's 1889).
1. ASHTON KEYNES. School 1870 ("truly horrible Gothic" P)   WRO
2. LYDIARD TREGOZE. Additions to school 1872   WRO
3. WROUGHTON. Lower Wroughton. Infants School 1872   WRO

LAVERS & BARRAUD stained glass makers
1. BLUNDSON ST ANDREW. St Andrews Church, E & SE windows to Butterfield's design 1868   P JO
2. SOUTH NEWTON. St Andrews Church chancel   JO

LAWRENCE architect fl 1734
1. DEVIZES. Wine Street/St Johns Street. New Hall (or Cheese Hall) 1734-5 C or 1750-2 P (Lawrence's design was made c1734. But is the present building of 1750, his?)

LAWSON, F.A architect of Stroud
1. NORTON. Manor House, alterations 1901 onwards   P

LEACHMAN, John architect 1795-
Pupil of James Medland of Southwark.   C
1. WARMINSTER. Sambourne. Christ Church 1830-12   P
2. CORSLEY. St Margaret's Church 1832-3. Altered by F.W.Hunt in 1891 C

LEDBURY, J.W architect
Practice at 63 Fore Street, Trowbridge (Kelly's 1923)

LEE, James iron church builder of 6 Buckingham Street, Manchester
1. SOUTHWICK. Iron church 1901 (now at Brokerswood)   WRO

LEMON & BUZZARD architects
Practice at 29 Market Place, Salisbury (Kelly's 1907, 1911)

LIGHT, William architect or builder
1. HULLAVINGTON. Additions to school 1870   WRO

LIGHT & SMITH builders of Chippenham
1. CALNE WITHOUT. Pewsham House 1892 (C.E.Ponting architect)

LINNELL, John wood carver
1. CALNE WITHOUT. Bowood House, chimney pieces c1760-1765   WAM 41

LISTER, James architect of Salisbury
1. STOCKTON. Parsonage 1891   WRO

LITTLE, Robert surveyor
1. WOOTTON BASSETT. Additions to Infants School 1868   WRO

LIVESAY, Augustus F architect of Portsmouth
1. TROWBRIDGE. Holy Trinity Church 1838   P

LLEWELLINS & JAMES of Bristol
1. BRADFORD-ON-AVON. Holy Trinity Church. Tenor bell c1890   PMS
2. GREAT SOMERFORD. Church. Bell re-cast 1897   VCH
3. CHIPPENHAM. Union Road, Slades Brewery. Fittings including white glazed
   bricks to cooling rooms c1898 (The Pictorial Record)

LONG, Albert Frank architect
Practice at 1 Ash Walk, Warminster (Kelly's 1889, 1895; at 21 East Street
(Kelly's 1899); as LONG & GLASS, at 53 Market Place (Kelly's 1907 - 1927 at
least).

LONG, James builder of Bradford-on-Avon 1820-1911
Son of William Long, mason.   RM and MA
1. BRADFORD-on-AVON. The Hall, restoration 1848   RM and MA
2. BRADFORD-on-AVON. Town Hall 1855 (co-builder with J.Spender) RM and MA
3. BRADFORD-on-AVON. Belcombe Place. Belcombe Lodge. Empty plot owned by
   him in 1864. House built between 1870-80   PMS
4. BRADFORD-on-AVON. Church Street, North Wilts Bank (now Lloyds Bank),
   designed and built it c1871   HF
5. DEVIZES. 'Much building work' after 1871   RM and MA

LONG, J & SONS builders of Bath
1. TROWBRIDGE. County Hall 1938-40 (architect P.D.Hepworth)   KR

LONG, Stanley Howard builder of Station Approach, Frome Road, Bradford-on-Avon
Leased part of Newtown Brewery as builder's yard from about 1926 and bought
the brewery premises in 1955. Trading as **Stanley H. LONG & SON** in 1957. Son
Michael Berkeley Long subsequently took over yard.   PMS

LONG, William builder of Albert Terrace, Trowbridge Road, Bradford-on-Avon
Born between 1820 and 1835. Son of William and brother of James. Bought
surplus land after construction of railway at Bradford.   RM and MA
1. BRADFORD-ON-AVON. Silver Street. New Mills (c1845?)   YP
2. BRADFORD-on-AVON. Victoria Terrace from 1859, fine imposing row   RM and
   MA

3. BRADFORD-on-AVON. Albert Terrace and Cottage 1863   RM and MA
4. BRADFORD-ON-AVON. Church Street. Abbey Mills 1875 (architect R.Gane) YP

LOVELL, Thomas master mason of Trowbridge Fl. c1480-1500
1. STEEPLE ASHTON. St Mary's Church. N aisle and possibly the whole.   KR

LOWDER, John architect of Bath
Surveyor to City of Bath 1819-22 and built Holy Trinity Church, Bath in Gothic style. NJ
1. BISHOPSTONE S. Rectory, 1815   VCH

LUSH, Edmund carpenter and joiner
1. SALISBURY. General Infirmary 1768 and Clerk of the Works for project.
   RCHM

LUTYENS, Sir Edwin Landseer architect 1869-1944
Born in London. Designed country houses, churches and other public buildings in Arts and Crafts, Baroque and Classical styles.   LB CH
1. MILTON LILBOURNE. St Peters Church. Restoration work 1924 (with
   H. Brakspear)   WRO
2. OAKSEY. Hill Farm, NW wing c1934 (E.J.T.Lutyens?)   VCH

LYONS,
1. SALISBURY. The Close, Arundells   RCHM

MACFARLANE & Co ironfounders? of Glasgow
1. SALISBURY. St Edmunds Church Street, Wesleyan Chapel. Open cast-iron
   gallery front   RCHM

MACKENZIE, David architect
Practice in Bank Street, Melksham (Kelly's 1889)

MACPHAIL, L architect
1. SEDGEHILL AND SEMLEY. Semley. Teacher's house 1865   WRO

MacVICAR ANDERSON J architect
1. CORSHAM. Hartham Park. Additions to house 1888   P

MACY, Phillip stone engraver of Chilmark
Died 24.2.1711 aged 45.

MAGGS AND HINDLEY engineers
1. STOURTON WITH GASPER. Stourhead, iron bridge 1860   P

MAJOR, George builder of Swindon
Stone and marble mason of Newport Street in 1842. ETD
1. SWINDON. Market Square, Town Hall 1853   MJ

MANNERS, George Philips architect of Bath 1789-1866
City Architect of Bath in early C19. In partnership with C. Harcourt Masters for a time. C Built several churches in Bath. In partnership with Gill (see below).   JO
1. LIMPLEY STOKE. Viaduct 1834 (D.Aust builder)   JO
2. BRADFORD-on-AVON. Bearfield, Christ Church 1841 (the chancel is

later - G.Scott 1878; other alterations J.O.Scott 1878)   PC
3. DILTON MARSH. Dilton Court 1842 (in association with John Peniston.
   D.Aust, builder)   Owner's plans
4. BROMHAM. St Nicholas Church. Restoration work 1843   WRO
5. TROWBRIDGE. Church Street, National Schools and house 1845 (Furniture
   showroom. JO)   WRO P
6. BRADFORD-on-AVON. Christ Church Schools and school house 1848
   (presumably by Manners)   JO

MANNERS & GILL architects of Bath
(See G.P. Manners)
1. KINGSTON DEVERILL. St Mary's Church. Nave, S aisle and chancel 1846
   P
2. TROWBRIDGE. St James Church. Restoration 1847-8   P
3. TROWBRIDGE. Hilperton Road. Highfield 1859 KR (P says by Goodridge)
4. TROWBRIDGE. Hilperton Road. Rodwell Hall 1859   KR
5. BRADFORD-on-AVON. Holy Trinity Church. Part of N aisle arcade 1864
   P WRO

MANNING, Joseph architect of Corsham
1. STANTON ST QUINTIN. Upper Stanton St Quintin Farm. Possibly the Mr
   Manning who was superintendant of building work there in 1852-54 EG
2. BISHOPS CANNINGS. Parsonage 1863   WRO

MANNINGS, George & SONS builders of Claverton Down, Bath
C. Beazer worked for Mannings in 1936 as mason and broke away to set up own
firm. VH2
1. COLERNE. Vale Court. Additions and renovations for Capt. R.D.Wills 1936
   VH2

MANNS family bricklayers and builders of Pewsey
George (aged 30), Mary (aged 25) and 3 children moved from London to Pewsey
c1825 and moved to 56 Wilcot Road between 1825 and 1839. In 1841 they had
10 children and Mary was a bricklayer, George a labourer. 7 of them
including George and the eldest girls were bricklayers by 1851. In 1861 son
James was married and a bricklayer off High Street. Thomas, 4th son, was 40
in 1875 and still at No. 56. A Mrs Manns, builder, there in 1894-1903
(Gillmans directory) may be Thomas's widow.   WBR JNH
1. PEWSEY. 56 Wilcot Road. Family owned. Ornate brickwork displaying their
   craft c1830-1860   WBR

MANTELL, Edward Walter architect of Swindon
1. BLUNSDON ST ANDREW. Blunsdon Abbey: design exhib. RA 1858 & 1860, built
   for C. de Windt.   JO
2. LYDIARD TREGOZE. School and house 1859. Design exhib. RA 1860 WRO JO
3. PURTON. School and house 1859   WRO
4. SWINDON. Rodbourne Cheyney, Vicarage 1863   JO

MARCHANT, Robert
1. BRADFORD-on-AVON. Holy Trinity Church. Restoration work 1928 WRO

MARSH, stone mason of Tisbury
Working on bridges at Codford and Fonthill 1829-30.   PL

MARTIN, stone tiler
1. SEMINGTON. Whaddon House and stable. Work 1673 and 1677   LP

MARTIN, A.C architect of Frome, Somerset
1. SOUTH WRAXALL. Manor House. Restoration 1900-02   VCH
2. WEST LAVINGTON. All Saints Church. Restoration work 1909 WRO
3. MARKET LAVINGTON. St Mary's Church. Restoration work 1910 WRO
4. LITTLE CHEVERELL. Hawkswell House 1914-20   P
5. WEST LAVINGTON. Dauntsey's School. Martin was architect up to
   c1920   P

MASLEN, Levi bricklayer and builder of 13 Bridewell Street, Devizes
Builder's yard to rear of house. PMS Firm est. 1890. Contractors to WCC in
1932 and other public bodies. **L. MASLEN & SONS** by 1916. One son was
William Jesse. WG
1. DEVIZES. The Arches, 14A Bridewell Street. On site of Wellington Court.
   Rebuilt to give access to his workshops 1905-6   WBR
2. ROUNDWAY. Wilts County Mental Hospital. Rest block for Nurses, by 1932
   (Gillman's 1932)
3. DEVIZES. Bricksteed Avenue. Working Men's Dwellings, by 1932 (Gillman's
   1932)
4. DEVIZES. Sheep Street. Flats on site of condemned housing 1950s? WBR

MASON, G builder
1. ALDBOURNE. School 1858 (architect Butterfield)   WRO

MASON, Paul freemason of Bradford-on-Avon
Leasing cottage at Newtown quarry 1607.   MP

MASTERS, William Arthur Harvey architect 1876 - ?
South Lodge, Stanton Fitzwarren. Practice at 1 Regent Circus, Swindon
(Kelly's 1907): 42 Cricklade Street (Kelly's 1911-1917 at least). Photo. in
D.
1. SWINDON. Little London, Mission Room c1902-6   JO D
2. STRATTON ST MARGARET. Upper Stratton, St Paul's Church 1904 JO
3. SWINDON. Even Swindon. Summers Street, St Augustine's Church 1907 P
   "A successful design"- PW
4. SWINDON. Broad Street, St Luke's Church 1911-12 P "The most beautiful
   church in Swindon."   JB
5. STRATTON ST MARGARET. Upper Stratton, St Philip's Church   D
6. SOUTH MARSTON. Village Hall   D
7. HIGHWORTH. Redlands Court (house for James Arkell)   JO D
8. WROUGHTON. St John & St Helen's Church, reredos & cross c1920 as War
   Memorial   JO
9. WOOTTON BASSETT. St Bartholomew's Church. Restoration work 1921 WRO
10. HIGHWORTH. War Memorial Cross   JO

MATTHEWS, H.W
1. CHIPPENHAM. Little George Hotel c1905   JO

MEDLICOTT, W.B architect
Practice at 15 High Street, Devizes (Kelly's 1907-1915)
1. BROUGHTON GIFFORD. St Marys Church. Restoration work 1906 WRO

later - G.Scott 1878; other alterations J.O.Scott 1878)   PC
3. DILTON MARSH. Dilton Court 1842 (in association with John Peniston. D.Aust, builder)   Owner's plans
4. BROMHAM. St Nicholas Church. Restoration work 1843   WRO
5. TROWBRIDGE. Church Street, National Schools and house 1845 (Furniture showroom. JO)   WRO P
6. BRADFORD-on-AVON. Christ Church Schools and school house 1848 (presumably by Manners)   JO

MANNERS & GILL architects of Bath
(See G.P. Manners)
1. KINGSTON DEVERILL. St Mary's Church. Nave, S aisle and chancel 1846   P
2. TROWBRIDGE. St James Church. Restoration 1847-8   P
3. TROWBRIDGE. Hilperton Road. Highfield 1859 KR (P says by Goodridge)
4. TROWBRIDGE. Hilperton Road. Rodwell Hall 1859   KR
5. BRADFORD-on-AVON. Holy Trinity Church. Part of N aisle arcade 1864   P WRO

MANNING, Joseph architect of Corsham
1. STANTON ST QUINTIN. Upper Stanton St Quintin Farm. Possibly the Mr Manning who was superintendant of building work there in 1852-54 EG
2. BISHOPS CANNINGS. Parsonage 1863   WRO

MANNINGS, George & SONS builders of Claverton Down, Bath
C. Beazer worked for Mannings in 1936 as mason and broke away to set up own firm. VH2
1. COLERNE. Vale Court. Additions and renovations for Capt. R.D.Wills 1936   VH2

MANNS family bricklayers and builders of Pewsey
George (aged 30), Mary (aged 25) and 3 children moved from London to Pewsey c1825 and moved to 56 Wilcot Road between 1825 and 1839. In 1841 they had 10 children and Mary was a bricklayer, George a labourer. 7 of them including George and the eldest girls were bricklayers by 1851. In 1861 son James was married and a bricklayer off High Street. Thomas, 4th son, was 40 in 1875 and still at No. 56. A Mrs Manns, builder, there in 1894-1903 (Gillmans directory) may be Thomas's widow.   WBR JNH
1. PEWSEY. 56 Wilcot Road. Family owned. Ornate brickwork displaying their craft c1830-1860   WDR

MANTELL, Edward Walter architect of Swindon
1. BLUNSDON ST ANDREW. Blunsdon Abbey: design exhib. RA 1858 & 1860, built for C. de Windt.   JO
2. LYDIARD TREGOZE. School and house 1859. Design exhib. RA 1860   WRO JO
3. PURTON. School and house 1859   WRO
4. SWINDON. Rodbourne Cheyney, Vicarage 1863   JO

MARCHANT, Robert
1. BRADFORD-on-AVON. Holy Trinity Church. Restoration work 1928   WRO

MARSH, stone mason of Tisbury
Working on bridges at Codford and Fonthill 1829-30.   PL

MARTIN, stone tiler
1. SEMINGTON. Whaddon House and stable. Work 1673 and 1677   LP

MARTIN, A.C architect of Frome, Somerset
1. SOUTH WRAXALL. Manor House. Restoration 1900-02   VCH
2. WEST LAVINGTON. All Saints Church. Restoration work 1909   WRO
3. MARKET LAVINGTON. St Mary's Church. Restoration work 1910 WRO
4. LITTLE CHEVERELL. Hawkswell House 1914-20   P
5. WEST LAVINGTON. Dauntsey's School. Martin was architect up to c1920   P

MASLEN, Levi bricklayer and builder of 13 Bridewell Street, Devizes
Builder's yard to rear of house. PMS Firm est. 1890. Contractors to WCC in 1932 and other public bodies. **L. MASLEN & SONS** by 1916. One son was William Jesse. WG
1. DEVIZES. The Arches, 14A Bridewell Street. On site of Wellington Court. Rebuilt to give access to his workshops 1905-6   WBR
2. ROUNDWAY. Wilts County Mental Hospital. Rest block for Nurses, by 1932 (Gillman's 1932)
3. DEVIZES. Bricksteed Avenue. Working Men's Dwellings, by 1932 (Gillman's 1932)
4. DEVIZES. Sheep Street. Flats on site of condemned housing 1950s? WBR

MASON, G builder
1. ALDBOURNE. School 1858 (architect Butterfield)   WRO

MASON, Paul freemason of Bradford-on-Avon
Leasing cottage at Newtown quarry 1607.   MP

MASTERS, William Arthur Harvey architect 1876 - ?
South Lodge, Stanton Fitzwarren. Practice at 1 Regent Circus, Swindon (Kelly's 1907): 42 Cricklade Street (Kelly's 1911-1917 at least). Photo. in D.
1. SWINDON. Little London, Mission Room c1902-6   JO D
2. STRATTON ST MARGARET. Upper Stratton, St Paul's Church   1904 JO
3. SWINDON. Even Swindon. Summers Street, St Augustine's Church 1907 P "A successful design"- PW
4. SWINDON. Broad Street, St Luke's Church 1911-12 P "The most beautiful church in Swindon."   JB
5. STRATTON ST MARGARET. Upper Stratton, St Philip's Church   D
6. SOUTH MARSTON. Village Hall   D
7. HIGHWORTH. Redlands Court (house for James Arkell)   JO D
8. WROUGHTON. St John & St Helen's Church, reredos & cross c1920 as War Memorial   JO
9. WOOTTON BASSETT. St Bartholomew's Church. Restoration work 1921   WRO
10.HIGHWORTH. War Memorial Cross   JO

MATTHEWS, H.W
1. CHIPPENHAM. Little George Hotel c1905   JO

MEDLICOTT, W.B architect
Practice at 15 High Street, Devizes (Kelly's 1907-1915)
1. BROUGHTON GIFFORD. St Marys Church. Restoration work 1906 WRO

2. WORTON. Christ Church. Restoration work 1908   WRO
3. SEEND. Lodge at Cleeve House?   JO

MERRICK, Henry architect of Bradford-on-Avon
1. BRADFORD-on-AVON. Parsonage (alterations?) 1880   WRO

MESSENGER, Henry architect
Practice at The Close Gatehouse, Salisbury (Kelly's 1911-1927 at least)
1. SALISBURY. East Harnham. St George's Church. Restoration work 1915   WRO
2. WESTWOOD. St Mary's Church. Restoration work 1924   WRO
3. BRIXTON DEVERILL. St Michael's Church. Restoration work 1928   WRO
4. WORTON. Christ Church. Restoration work 1929   WRO
5. NETHERAVON. All Saints Church. Restoration work 1931   WRO
6. BRATTON. St James' Church. Restoration work 1932   WRO
7. ENFORD. All Saints & St Margaret's Church. Restoration work 1932   WRO
8. EVERLEIGH. St Peter's Church. Restoration work 1933   WRO
9. URCHFONT. St Michael's Church. Restoration work 1934   WRO
10. WOODBOROUGH. St Mary Magdalene Church. Restoration work 1935-6
11. OGBOURNE ST GEORGE. Restoration work 1936   WRO
12. LITTLE BEDWYN. St Michael's Church. Restoration work 1936   WRO
13. FROXFIELD. All Saints Church. Restoration work 1937   WRO
14. HAM. All Saints Church. Restoration work 1937   WRO
15. BAYDON. St Nicholas Church. Restoration work 1937   WRO
16. CHIRTON. St John's Church. Restoration work 1938   WRO
17. WOOTTON RIVERS. St Andrew's Church. Restoration work 1938   WRO
18. GREAT BEDWYN. St Mary's Church. Restoration work 1938   WRO

MESSENGER, conservatory builders of Loughborough
1. SWINDON. King Edward's Place c1910   JO
2. SWINDON. Kingsdown House (supplied to T. Arkell)   JO
3. TROWBRIDGE. Roundstone Street, Polebarn House   JO

METHUEN, Hon. Anthony P architect of Corsham
1. MONKTON FARLEIGH. St Peter's Church. Restoration work 1933   WRO
2. CORSHAM. Hungerford Almshouses. Improvements, bathrooms
   etc. 1938   WRO
3. CALNE WITHOUT. Whetham House. Alterations.   ifo

MICKLETHWAITE, John Thomas architect 1843-1906
Pupil of George Gilbert Scott. In partnership with Somers Clarke 1876-92.
Surveyor to the Dean and Chapter of Westminster Abbey, 1898.   DNB
1. INGLESHAM. St John the Baptist. Restoration work 1889-9 (for SPAB in
   assoc. with Wm. Morris). P

MILEHAM, G.S architect of Cuffley, Herts
1. SALISBURY. St Thomas's Church. Restoration work 1926   WRO

MILES, Thomas B builder of Shaftesbury, Dorset
1. TISBURY. Workhouse 1869 (architect Creeke)   RG
2. ALVEDISTON. School 1872   WRO
3. SUTTON MANDEVILLE. School and house   WRO

MILLAR, Joseph architect and surveyor of Back Street, Trowbridge
1822 3. ETD

MILLER, builder of Seagry
1. KINGTON LANGLEY. St Peter's Church 1855-6   M

MILLER, Sanderson amateur architect 1717-1780
Gothic style. His mason was William Hitchcock.   C
1. LACOCK. Abbey, gothicized by Miller 1754-5; also a Gothic
   Gateway   PC

MILLINGTON, William artist of Trowbridge
May have worked with William Smith, architect (qv).
1. SOUTHWICK. Primary School. Elevations said to be from design by
   Millington 1860s   ML

MINTY & GODWIN glaziers, painters and plumbers
1. SALISBURY. General Infirmary 1768   RCHM

MITCHELL, Edward carpenter of Bradford-on-Avon
1. BRADFORD-ON-AVON. Newtown quarry. Erected own house 1607   MP

MITCHELL, John architect of Pewsey
1. PEWSEY. National School. Plan of room 1860   WRO
2. WOOTTON RIVERS. School 1864   WRO

MIZEN, William mason
1. DAUNTSEY. Rectory 1829-33   DR

MOFFATT, W.J architect
Partner of G.G.Scott (see also SCOTT and MOFFATT)
1. TEFFONT. Teffont Evias Vicarage 1841   WRO   JO

MONDEY, Edward architect of Dorchester, Dorset
1. SALISBURY. Fisherton Anger. Parsonage 1865   WRO

MONEY, John architect of Donnington
1. MARLBOROUGH. St Mary's Parsonage 1839   WRO
2. WILCOT. Parsonage 1842   JO   WRO

MONEY, Thomas architect of Donnington
1. MELKSHAM WITHOUT. Shaw, Parsonage 1827   WRO

MOODY, Lemuel builder of Trowbridge
Three generations of this family were builders   ML
1. SOUTHWICK. Primary School 1860s   ML

MOORE, Temple Lushington architect 1856-1920
Pupil of George Gilbert Scott jnr, Gothic style.   DNB
1. SWINDON. St Mark's Church. Lengthening of, and alterations to,
   chancel 1897. (Church designed by Scott & Moffatt)   P

MORGAN, H.T architect of London   WWA
1. BRADFORD-ON-AVON. Trowbridge Road, Council Estate? c1920   JO

MORGAN, J.L architect
Swindon Borough Architect.

- 64 -

1. SWINDON. Walcot East. Sussex Square (pedestrian shopping square) 1958 P
2. SWINDON. Penhill. Tower blocks 1960s (with M. de St Croix)   P
3. SWINDON. Courts of Justice and Magistrates' Courts 1963-4   P

MORLIDGE, John
1. EVERLEIGH. St Peter (the village was rebuilt on a new site 1810-11) P

MORRIS, Frank architect
Son of County Surveyor, Berks.
1. SWINDON. Regent Street, McIlroys shop   JB

MORRIS, Roger carpenter, architect and engineer
Principal engineer to the Board of Ordnance 1695-1749. An exponent of Campbell's Palladianism. Also tendency to medievalising. Most of his work was done in association with the 9th Earl of Pembroke. Works include Marble Hall, Twickenham.   F, H & P RCHM
1. DOWNTON. Standlynch. Trafalgar House 1733 (the centre of the present house)   P
2. WILTON. Wilton House. State bedroom (Colonnade Room). Ionic columns probably added c1735 by Morris   P
3. WILTON. Wilton House. Palladian Bridge 1736-7 (design by the 9th Earl of Pembroke, with Morris as architect)   P RCHM
4. WILTON. House. Porter's Lodge (attribution), now gone.   RCHM
5. SWINDON. Lydiard Tregoze. Lydiard Park. 1743 onwards (attribution by C.Hussey).   P RCHM
6. ODSTOCK. Longford Castle. Alterations to facade c1750 (or may have been designed by Theodore Jacobsen).   P

MORRIS, William designer and craftsman 1834-1896
Educated at Marlborough and Oxford. Architectural student in Street's office. Head of a group making stained glass and furniture. LB Lived at Kelmscott Manor, Oxon. Founded the SPAB in 1877.
1. MALMESBURY WITHOUT. Rodbourne. Holy Rood. Stained glass E window. Designed by Ford Madox Brown and D.G.Rosetti. Made by Morris & Co. 1865. "Early work"- P VCH
2. BROMHAM. St Nicholas Church. E window 1870 (figure work from Burne-Jones cartoons).   P
3. SOPWORTH. St Mary. Stained glass by Morris and Co c1870   P
4. SALISBURY. Cathedral. Stained glass c1875-80   P
5. INGLESHAM. St John the Baptist's Church. "This church was repaired in 1888-9 through the energy and with the help of Wm. Morris who loved it"- notice in church. (Architect J.T.Micklethwaite).   P

MORRISH, W.J.M architect of Gillingham
1. CALNE. St Mary's Church. Restoration work 1907   WRO

MOULTON & ATKINSON architects of Salisbury
1. STOURTON WITH GASPER. Stourhead. The wings of the house 1792-1804 P

MOUNT, G.E architect of Salisbury
1. SHREWTON. Maddington. St Mary's Church. Restoration work   WRO

- 65 -

MOWBRAY, GREEN & HOLLIER architects
1. TISBURY. St John's Church. Restoration work 1922   WRO
2. WARMINSTER. St Denys Church. Restoration work 1927   WRO

MULLENS (or MULLINS or MULLINGS), Benoni architect and builder
Cabinet maker of Northgate Street in 1842. ETD
1. DEVIZES. British School 1860   WRO
2. DEVIZES. Northgate Street, Congregational Chapel, lecture hall and
   schoolroom 1868-9 P Designed and built by him. RCHM

MUNDY, Herbert architect and surveyor of Westbourne Gardens, Trowbridge
1851-
Senior partner of **FOLEY, SON and MUNDY** (qv), auctioneers, surveyors and
land agents established 1845. D Valuer for WCC, saleroom in Manvers
Street (Trowbridge Directory 1901). See also FOLEY.
1. HEYWOOD. Parsonage 1898   WRO

NAISH, Giles of Salisbury
1. SALISBURY. 47 Winchester Street, Three Cups Inn. Leased 1671 from city
   with covenant to pull down and rebuild within 2 years. Lease passed to
   Thomas Naish (qv) so G. Naish may have been a builder. RCHM (Salisbury
   vol 1)

NAISH, Thomas of Salisbury, fl. 1694-1705
Clerk of Works of Cathedral.   RCHM (Salisbury vol 1 p139)

NASH, John town-planner and architect 1752-1835
Best known for the layout of Regent's Park & Regent Street, London, 1811
onwards. Started Buckingham Palace, but dismissed after the death of
George IV in 1830.   F, H & P
1. CORSHAM. Corsham Court. Englargements 1800   P
2. CORSHAM. Corsham Court. Lake Cottage   P

NELSON, J.M architect
1. MARLBOROUGH. House (Castle Inn) now C House of Marlborough College.
   Conversion plans 1842   VCH
   (Report Marlborough Coll. Nat.Hist.Soc. 1957 No 98 p43-53)

NEWMAN, Dudley architect of London
1. REDLYNCH. St Birinus. Restoration work 1919   WRO

NEWMAN family thatchers of Christian Malford
Lived in cottage at The City, near Brights Farm. Three generations of
thatchers going back from one who died aged 86 in c1924. (N. Wilts Herald
16.10.1931)

NEWMAN, Richard mason of Bearfield, Bradford-on-Avon, -1850
Married and at Bearfield in 1813 (parish registers) and 1822 ETD. May have
built several cottages at Bearfield which he owned or possibly built by his
father William Newman. Occupying a garden quarry at Bearfield and one at
Winsley Road in 1841. GL
1. BRADFORD-ON-AVON. Belcombe Place. Nos 13 and 14 by 1828. Nos. 11 and 12
   1837-40. All built on an empty plot and let out.   PMS

- 66 -

NEWMAN, Richard builder of Bath Road, Bradford-on-Avon, -1885
Harrods Directory 1865. Son of the above. PMS

NEWTON, Sir Ernest architect 1856-1922
Founder member of the Art Workers' Guild in 1884. Art Nouveau designer of the Arts and Crafts Movement. Best known for Buller's Wood (Chislehurst), 1890; Chapel for the Sisters of Bethany, Lloyd Square, 1891; Redcourt (Haslemere) 1894, etc.   G
1. MARKET LAVINGTON. Clyffe Hall. Alterations (entrance motif of porch, Venetian window, & oval window over, perhaps top pediment) 1904   P
2. MARLBOROUGH. Priory House. W. block 1926 (after death, W.G.Newton?) VCH

NEWTON, Ernest architect of London
1. MANNINGFORD. Manningford Abbots. Church. Restoration work 1937   WRO

NEWTON, Professor William G architect
Marborough College architect.
1. MARLBOROUGH. College. Memorial Hall 1921-5. Elmhurst House, additions 1923. Science building 1933, one of first exposed aggregate reinforced concrete buildings in England (builder Rendell). R P Leaf Block 1936 VCH
2. WILCOT. Holy Cross Church. Restoration work 1937   WRO

NICHOLAS, James carpenter of Upavon
Asked by J. Peniston if he would repair Upavon Bridge in 1825. PL

NICHOLLS, Herbert Edward architect 1877-? of Goddard Avenue, Swindon
Articled to C.W.Evans of Southampton. Partnership - NICHOLLS & STOCKWELL, 25 Regent Circus, Swindon (Kelly's 1907). (See also STOCKWELL.)   D
1. SWINDON. Ferndale School (with Stockwell)   D

NICHOLS, George Benjamin architect of Birmingham
1. ALDERBURY. Workhouse 1879 (succeeded by Hall)   RG

NICHOLSON, Sir Charles architect
Follower of Bodley.   JO
1. WARMINSTER. St Boniface College, Chapel block/library 1927   P
2. POTTERNE. St Mary's Church. Restoration work 1936   WRO

NOYES, William architect
1. POULSHOT. Parsonage 1784   WRO

NOYES & GREEN, architects
1. AMESBURY. St Mary's Church. Restoration work 1938   WRO

OAKFORD, Charles builder
1. CALNE WITHOUT. Bowood House. Work c1770-1775   WAM 41

O'CONNOR, M and A of Ireland and London
Awarded Gold Medal at 1862 Exhibition.   JO
1. BRADFORD-ON-AVON. Holy Trinity Church, E window glass 1857   JO
2. SALISBURY. Cathedral, NE windows 1859   P
3. CALNE. St Mary's. S Chapel E window 1866   P JO

ODBER, John builder
1. STOURTON WITH GASPER. Stourhead Pantheon 1753, 'fairly certain attribution' KAR

OLDRIEVE, W.T architect
1. SALISBURY. Castle Street Post Office 1907 P

OLIVER, C.Bryan architect of Bath
1. CALNE. Town Hall 1884-6 P

THE TOWN HALL.

Town Hall, Calne (from A.E.W. Marsh 'History of Calne' 1904)

OLIVER, E Keene architect
Practice at The Manor House, Manningford Abbotts (Kelly's 1911, 1915)
1. UPAVON. St Mary's Church 1930 WRO
2. BURBAGE. All Saints Church. Restoration work 1933 WRO

OSBORNE AND SONS builders of Corsham
Monumental masons, builders and contractors. Established 1775 and continued to death of Mr Bert Osborne in 1960s. Were for many years the largest builders in the area. Much of work in larger houses of district. (See 1933 photograph in Wiltshire Times 15.3.1985).

OSBORNE, William Robert architect 1878-?
Born Corsham. Lived at Whiteway, Broadtown. Practice at 37 Regent Circus, Swindon (Kelly's 1907). Worked both with W.H.Read of Swindon and H. Brakspear. D
1. SWINDON. Baptist Tabernacle, Victoria Hospital, Clifton Street

Schools, Rodbourne Cheyney Primitive Methodist Church, County Ground Hotel and Even Swindon Hotel etc (all with W.H.Read) D

OSGOOD, Ivor carpenter and barn builder
Buried at Cholderton. Universal British Directory of 1793-8 also gives Francis Osgood, carpenter, John Osgood, carpenter and joiner and William Osgood, carpenter and joiner, all under Amesbury. ETD
1. AMESBURY. Countess Farm. Staddle barn 1772 and small barn undated. RCHM

OSMOND, W builder
1. SALISBURY. Poultry Cross, upper part 1852-4 (designed by O.B.Carter) RCHM

OSMOND, Walter William architect
Practice at Clarendon Park (Kelly's 1899)

OVERTON, Samuel architect and builder
Practice at Forest Hill, Marlborough (Kelly's 1875)
1. MARLBOROUGH. St Mary's. School, additions 1859 WRO
2. COLLINGBOURNE DUCIS. School and house 1861 WRO VCH
3. LITTLE BEDWYN. Parsonage 1873 WRO

OVERTON, Thomas Collins architect
1. DEVIZES. Bath Road. A "summer house" for James Maynard c1766 (very close to the present site of Braeside) VCH

PACE, Richard builder & architect of Lechlade, Glos c1760-1838
1. CLYFFE PYPARD. Bushton, Woodhill Park. SE range added 1804 P
2. CRICKLADE (?). Broadlease 1808 for E. Lovden C
4. BISHOPSTONE N. Hinton Rectory 1810 C
5. WROUGHTON. Parsonage 1828 WRO

PAINE, James architect c1716-1789
A country house architect, working in Burlington's neo-Palladian style, but he was eclipsed eventually by Robert Adam. F, H & P
1. TISBURY. New Wardour Castle. Designed by Paine for 8th Lord Arundell. Begun 1770, completed 1776 P
2. TISBURY. New Wardour Castle. Chapel 1776 (lengthened by Soane 1789-90) P

PARKER & UNWIN town planners ie Barry Parker 1867-1941, and Sir Raymond Unwin 1863-1940
The translators of Ebenezer Howard's "garden city" ideals into practice. Letchworth begun 1903: Hampstead Garden Suburb begun 1907: Wythenshawe (Manchester) begun 1927.
1. SWINDON. Pinehurst Estate 1919 (original part) P

PARKER, Charles architect 1800-1881
Pupil of Sir J. Wyatville. Founder fellow of RIBA. C
1. STOURTON WITH GASPER. Stourhead. Portico 1841 P

PARKER, Robert John architect
Practice at Market Place, Melksham (Kelly's 1895)

PARKER-PEARSON, George
(see G.P Pearson)

PARSONS, George carpenter and joiner of Amesbury
Universal British Directory 1793-8. ETD
1. WILTON. Wilton House. New chimneys, sash windows and door adjustments
   1781   RCHM

PAULL & BONELLA architects of Manchester and London
Leading chapel architects.
1. TROWBRIDGE. Church Street. The Tabernacle (Congregational) 1882   P

PEACOCK, Kenneth J.R architect
Of **Louis de SOISSONS & PARTNERS.**
1. LACOCK. Bowden Park. Work to replace Victorian additions 1950s   P

PEARCE, Alexander & SON builder
Established 1919. Of 88 Crane Street, Salisbury in 1969.

PEARCE, George and family thatchers of Wilcot
Operating from c1900 onwards, thatching many houses in Wilcot and
elsewhere. HTV programme

PEARSALL, Thomas iron founder of Willsbridge, Bitton, Glos
1. CALNE WITHOUT. Blacklands Mill. Stable roof. 1811 patent. One of 5 which
   Pearsall constructed of thin 'hoop iron' to his own patent system and
   described in trade document of 1812.   JO

PEARSON, John Loughborough architect 1817-1897
Essentially a church architect and a Gothicist, but also designed country
houses, e.g. Quarr Woode, Glos, 1857 (Gothic) and Westwood, Sydenham, 1881,
which is French Renaissance. Studied under Bonomi, Salvin & Hardwick.
Best churches 1870s - 1880s, including St Augustine, Kilburn Park (1870-
1880) and Truro Cathedral (1879-1910).   F,H & P
Calm proportion and simplicity of line.   JO
1. CHARLTON S. St Peter's Church 1858   P
2. LUDGERSHALL. St James Church. S chapel & seats 1858-9, chancel 1874 JO
   Restoration early 1970s VCH
3. NORTH NEWNTON. St James Church. Restoration work 1862-3   JO
4. KINGTON ST MICHAEL. Kington House 1863 (attributed)   JO
5. CHUTE FOREST. Chute Lodge repairs & alts. 1864-71   JO
6. IDMISTON. All Saints Church restoration 1865-7   P
7. SUTTON VENY. St John The Evangelist Church 1866-8   P
8. CHUTE. St Nicholas Church 1869-72   P
9. SUTTON VENY. Schools 1872-3   JO
10. CHUTE FOREST. St Mary's Church 1875   P
11. MILTON LILBOURNE. St Peter's Church. Restoration 1875   P
12. IDMISTON. Porton, St Nicholas Church 1876-7   P
13. MANNINGFORD. Manningford Bruce. St Peter's Church, restoration 1882 P
    (A. Quiney,'J.L. Pearson', 1979, pp264-5)   JO
14. TIDWORTH. Holy Trinity Church. Restoration 1882   VCH
15. CALNE. St Mary's Church. Restoration work 1890   WRO

PEARSON, George Parker architect also styled PARKER-PEARSON, George
Practice at Aspenden House, Silverless Street, Marlborough (Kelly's 1915);
by 1923 moved to Grittleton, then called himself Parker-Pearson.

PEDLEY AND SMITH builders of Highworth
William Pedley and Thomas Smith (qv).
1. STRATTON ST MARGARET. Highworth and Swindon Workhouse 1846   RG

PEDLEY, W architect
1. HIGHWORTH. Sevenhampton. St Andrew 1864   P

PENISTON, Henry architect of Salisbury 1832-1911
Practice at De Vaux Lodge, 9 De Vaux Place, Salisbury (Kelly's 1875-1899).
Grandson of John Peniston and son of John Michael. PL
1. SALISBURY. Devizes Road, Police Station 1858 (Building News 1858 170)   JO

PENISTON, John architect of Salisbury c1778-1848
Died 22.6.1848 aged 69 (tomb at St Osmund's R.C. Church). County Surveyor of roads, bridges etc. for Quarter Sessions and was concerned with many parsonage dilapidations. The Peniston papers relating to the practice of the three Penistons are at WRO. PL PMS JF
1. POULSHOT. Parsonage 1823   WRO
2. BROMHAM. Rowdeford House, alterations for Wadham Locke (J.B.White builder) 1825   PL
3. CHIRTON. Conock Manor, proposed alterations 1825   PL
4. LUDGERSHALL. House of Charles Millet, extensive repairs, new coach house, wash-house and pantry 1825   PL
5. MARKET LAVINGTON. Clyffe Hall. Proposed alteration to stables 1825   PL
6. WYLYE. New rectory 1827   WRO
7. CLARENDON PARK. Clarendon House. Alterations 1829   PL
8. EAST KNOYLE. St Mary's Church. Enlargement 1829 (a gallery)   C
   (Extension of N transept into an aisle.)   VCH
9. SALISBURY. De Vaux Place - terrace c1830 'probably by John Peniston', plans at WRO   RCHM
10. SALISBURY. Bell and Crown Inn, stables 1831   JF
11. DEVIZES. St James' Church. Main body of the church 1832   P
12. SALISBURY. Trinity Hospital, alterations 1832   JF
13. ALLINGTON. Boscombe, Queen Manor. New range 1832   VCH
14. SALISBURY. Castle Street, malt house, stables and store for Mr Cooper 1834   JF
15. SALISBURY. Three Swans Inn, alterations (bar, kitchen etc) 1834   JF
16. DILTON MARSH. Dilton Court 1842 (in association with G.P.Manners. David Aust of Bath, builder)   Owner's plans

PENISTON, John Michael architect of Salisbury 1807-1858
Son of J. Peniston. Worked for a while in office of Thos. Hopper (qv). PL

PENNING, William H builder
1. PEWSEY. National School and house 1861 (architect G.E.Street)   WRO

PETO, Harold architect
Former partner of Sir Ernest George.
1. WESTWOOD. Iford Manor, alterations after 1899   JO

PHILIP, John Birnie sculptor 1824-1875
Spent 8 months on the Albert Memorial, London. DNB
1. SALISBURY. Chapter House. Restoration 1856 (under Clutton) P
2. WILTON. St Mary and St Nicholas. Tomb of Herberts 1856 and 1861
   (designed by Wyatt) P

PHILLIPS, John architect and builder of Devizes Road, Swindon
His works at the Swindon Quarries.
1. SWINDON. Market Square, Corn Exchange c1866
2. PURTON. New Infants School Room 1873 WRO

PHIPPS, G.J (G.L VCH) architect
1. LEA AND CLEVERTON. Lea. St Giles Church. Additions (i.e. most of it)
   1879-80 P

PHIPPS, J.C architect of Bath 1835-1897
(Same as C.J.Phipps DOE and WRO?). Later famous as theatre architect in
London viz. Savoy, Lyric and Her Majesty's. JO
1. PEWSEY. Cemetery Chapel 1862 P
2. LEA AND CLEVERTON. School and house 1872 (as C.J.P, London) WRO
3. PEWSEY. National School. Additions 1873 (as C.J.P.London) WRO

PICTOR & SONS quarry owners, stone merchants and stone masons of Box
Harrods Directory 1865. Own firm may have built some of the fine houses
designed for their family in Box area. PMS

PIERCE, Edward painter
1. WILTON. Wilton House. Wall and ceiling paintings mid 17thC RCHM

PIKE, Charles & PARTNERS architects
1. SWINDON. Victoria Road. The College 1956-8 P

PILKINGTON, William architect 1758-1848
Pupil of Sir Robert Taylor. Employed as surveyor and architect by the Earl
of Radnor. C
1. SALISBURY. Guildhall 1788-95 (designed by Sir Robert Taylor; built by
   Pilkington, with some alterations) P
2. CHILTON FOLIAT. Chilton Lodge (actually in Berkshire) for John Pearse
   1800 (since altered by Sir A. W. Blomfield). C
3. CALNE. Wick Hill House, for Charles Pole 1812, with large alterations
   1815 C

PINCH, John the elder builder and architect of Bath c1770-1827
Surveyor to the Darlington estates. Designed terraces on Bathwick and
Lansdown Hills, Bath. Work in the 18thC tradition brought to a new height
of elegance and refinement. NJ
1. BISHOPSTROW. Bishopstrow House 1817 P

PINCH, John junior architect
1. GRITTLETON. Church of St Mary. Rebuilt S aisle 1836 JO
2. CHIPPENHAM. Old workhouse, alterations prepared by Mr Pinch, Surveyor
   of Bath. RG

PINCHARD, Biddulph architect
Practice at Staple Inn, Holborn.
1. ALDBOURNE. Upper Upham, Upham House. Additions 1913. P Restorations
   1909-1922 VCH
2. CHISELDON. Holy Cross Church. Restoration work 1934-5   WRO
3. BRATTON. St James Church. Restoration work 1937   WRO

PINNEGAR, C.E architect or builder of Tytherton, Chippenham
1. LYNEHAM. School and house 1862   WRO

PINNEGAR, J builder
1. KINGTON LANGLEY. Union Chapel 1835   RCHM

PINFOLD, C.G architect
1. CHIPPENHAM. Post Office 1959   P

PIPER, T & W builders of London
1. ROUNDWAY. County Lunatic Asylum (later Roundway Hospital) 1849-1851
   (architect T.H.Wyatt)   RCHM

PITT, Sandford architect
Practice at Calstone Wellington, Calne Without (Kelly's 1923-?)

PITT, T.E architect
Practice in Heddington (Kelly's 1907-1915)

PLANK, John carpenter and builder of Long Street, Devizes   ETD (1830)
1. STANTON ST BERNARD. All Saints Church. New nave, chancel and vestry in
   Gothic style 1832   MM

PLEYDELL-BOUVERIE, Bertrand stone carver
1. STANTON ST QUINTIN. St Giles6s Church. Carved stone pulpit c1876 VCH

PLOWMAN, John architect of Oxford c1773-1843
1. DINTON. Baverstock. Parsonage 1826   WRO

POLLARD, architect of Frome, Somerset
1. DEVIZES. Market House 1835   P

PONTING, Charles Edwin architect
Practice at Lockeridge Cottage, West Overton (Kelly's 1875-1899), at Wye
House, 8 Barn Street, Marlborough (Kelly's 1907-1915). Diocesan architect.
Architect and surveyor for Marlborough College. DOE
1. SAVERNAKE. School and house 1871   WRO
2. WEST OVERTON. Lockeridge School c1872   VCH
3. WEST OVERTON. Overton and Fyfield School and house 1875   WRO
4. AVEBURY. School, additions to 1842 building, 1877   WRO
5. WEST OVERTON. St Michael's Church 1878   P
6. BROAD HINTON. St Peter Ad Vincula. Restoration 1879   P
7. LIDDINGTON. All Saints. Restoration 1880s ("a severe restoration") VCH
8. MERE. St Matthew's Church 1882   P
9. BISHOPS CANNINGS. St Mary's Church. Restoration work 1883   WRO
10. MARDEN. All Saints Church. Rebuilding 1885   P
11. POULSHOT. Wesleyan Chapel 1886 ("from designs by C.E.Ponting") VCH

12. EDINGTON. St Mary's, St Katherine and All Saints. Restoration 1887
    (1889-91 in VCH) P
13. STANTON ST QUINTIN. St Giles Church. Chancel 1888-9 VCH P
14. NETHERAVON. All Saints Church. Restoration 1888 VCH
15. PEWSEY. St John the Baptist. Restoration work 1889-90 P
16. LANGLEY BURRELL WITHOUT. Langley Burrell. Church chancel restored, 1890 JO
17. HOLT. St Katherine's Church 1891 P
18. CHISELDON. Holy Cross Church. Restoration 1892 VCH
19. CALNE WITHOUT. Pewsham, Old Derry Hill, Pewsham House 1892 JO
20. LATTON. Almshouses. Design exhibited at RA 1892 JO
21. ENFORD. All Saints Church. Restoration 1893 VCH
22. DONHEAD ST ANDREW. Parsonage 1893 WRO
23. BULFORD. Vicarage 1893 VCH
24. REDLYNCH. St Birinus Church 1894-6 P
25. SALISBURY. Bemerton. St John's Church, restoration work 1896 P
26. LEIGH, THE. All Saints Church. Chancel 1896 P
27. WEST LAVINGTON. Dauntsey's School 1895 (Ponting was the original architect). P
28. NORTH WRAXALL. Ford. St John the Evangelist Church 1897 P
29. STOCKTON. Parsonage 1898 WRO
30. MARKET LAVINGTON. Clyffe Hall c1898-1905 JO
31. URCHFONT. St Michael's Church. Restoration 1899-1900 VCH
32. SOUTHWICK. St Thomas's Church. 1899-1904 P
33. MARLBOROUGH. St Mary's Parsonage 1899. WRO
34. WOODFORD. All Saints Church. Restoration work 1899 WRO
35. MARLBOROUGH. Town Hall. Rebuilding 1901-2 VCH
36. MARLBOROUGH. Cross Lane, The Lodge c1902. Attributed DOE
37. MARLBOROUGH. Cross Lane, Hyde Cross 1902 for Marlborough College housemasters. Attributed DOE
38. MARLBOROUGH. Cross Lane, Clements Meadow 1902 for Marlborough housemasters. Attributed DOE
39. BULFORD. St Leonard's Church, restored 1902-11 VCH
40. BRINKWORTH. St Michael's Church, restoration 1902-3 VCH
41. DEVIZES. Town Hall. Restoration work 1903 R
42. FITTLETON. All Saints Church. Restoration of chancel and tower 1903 VCH
43. DEVIZES. 11-12 St John's Street. Reconstruction and conversion to offices and storerooms for F. Rendell (qv) 1903 R
44. DURNFORD. St Andrew's Church, nave and tower restored 1903-4 VCH
45. ALTON. Alton Barnes. St Mary's Church. Restoration 1904 P
46. SWINDON. Lydiard Tregoze. St Mary's Church. Restoration "at the beginning of the 20th century" VCH
47. MELKSHAM WITHOUT. Shaw. Christ Church. Rebuilding 1905 P
48. RUSHALL. St Matthew's Church. Restoration 1905 VCH
49. AMESBURY. St Mary and St Melor, restored 1905 (with D. Blow) VCH
50. DURNFORD. Vicarage, alterations and additions 1905-6 VCH
51. KEEVIL. St Leonard's Church. Restoration work WRO
52. MILSTON. St Mary's Church, vestry built 1906 and restoration 1913 WRO VCH
53. HULLAVINGTON. Church. N part restored and organ chamber built 1907 VCH
54. MARLBOROUGH. College. Gymnasium 1908 (incorporating parts of the Town Gaol) P
55. SALISBURY. St Martin's Church. Restoration work 1908, 1913, 1917 WRO
56. MELKSHAM. St Michael's Church. Restoration work 1909 WRO

57. BROMHAM. St Nicholas' Church. Restoration work 1909	WRO
58. ROWDE. St Matthew's Church. Restoration work 1910	WRO
59. SALISBURY. St Edmund's Church. Restoration work 1912	WRO
60. LITTLE BEDWYN. Parsonage 1912	WRO
61. CORSHAM. Gastard, Church of St John the Baptist, 1912	JO
62. ALDBOURNE. St Michael's Church. Restoration work 1914	WRO
63. OGBOURNE ST ANDREW. Restoration work 1914	WRO
64. NETHERAVON. All Saints Church. Restoration work 1917	WRO
65. BRADFORD-ON-AVON. Holy Trinity Church. Restoration work 1917	WRO
66. CLYFFE PYPARD. St Peter's Church. Restoration work 1919	WRO
67. MARLBOROUGH. St Mary's Church. Restoration work 1919, 1922	WRO
68. MARLBOROUGH. St Peter's Church Restoration work 1919	WRO
69. TIDWORTH. Holy Trinity Church. Restoration work 1919	WRO
70. POULSHOT. St Peter's Church. Restoration work 1920	WRO
71. WARMINSTER. Boreham, St John the Evangelist (Street, 1865, but baptistery at W end by Ponting 1925-6)	P

St John's Church, Ford, North Wraxall from Illustrated Building News 1898. As built the spire is shorter and the W. window larger.

PONTON, builder of Warminster
1. HOLT. Congregational Chapel 1880 (with Mr Brown, builder of Hilperton). (History of Holt United Reformed Church)

PONTON, Aaron builder? of Warminster
Saddler and harness maker of this name at the Market Place 1842. ETD
1. WARMINSTER. East Street, Yard House. Zinc chimney pot added to flue 1846 HA

POPE, John G architect
1. OGBOURNE ST ANDREW. School and house 1871   WRO

POPE, Richard Shackleton surveyor of Bristol
One of the District Surveyors for the City of Bristol 1831 until past 1850. C
1. WINSLEY. St Nicholas Church. Rebuilding 1841   P
2. BRADFORD-ON-AVON. Parsonage 1840   WRO   "Standard Tudor" JO

PORTER, T McEWAN architect
Trained at Arch. Asscn. and joined Imrie & Angell in London, in 1937. Warminster practice set up in 1950s, with Peter Wakefield, Peter Swann and Victor Harris (1956); Imrie, Porter & Wakefield of Market Place, Warminster. Diocesan Surveyor for Salisbury Archdeaconry. Retired 1948.

POTTER, Robert architect of Salisbury
Later of Brandt, Potter, Hare Parnership (qv) and later still, Potter and Partners and today this has become the Sarum Partnership.
1. DOWNTON. Charlton. All Saints. Restoration work 1934   WRO
2. SALISBURY. St Francis Church 1936-9. Consecrated 1940   P

POWELL, Isaac carpenter
Great grandson of John Button, builder of Calne.
1. CALNE WITHOUT. Bowood House. House carpenter there for nearly 50 years. WAM 41

POWELL, J George surveyor
The second WCC County Surveyor.
1. ROUNDWAY. Roundway Hospital, engineer's house 1909   RCHM
2. ROUNDWAY. Roundway Hospital, female staff mess room 1911 RCHM
3. TROWBRIDGE. Hill Street, County Offices (later Trowbridge Library) 1913   JO

POWELL, J & SONS glaziers of Whitefriars Works, London
1. TROWBRIDGE. St James Church. Glass, 1896 and 20thC   JO
2. WARMINSTER. Boreham, St John's Church. Mosaic tile decoration 1911-15   JO
3. MELKSHAM. Forest, St Andrew's Church. E Window 1875   JO
4. MERE. St Michael's Church. N aisle West window and S aisle West, window designed by H. Holiday 1865. 'Excellent, in the Pre-Raphaelite style' P   JO

POWELL, R Sidney architect of London
1. LONGBRIDGE DEVERILL. Hill Deverill. Church of the Assumption. Restoration work 1896   WRO

POWELL AND MOYA architects i.e A J Philip Powell b.1921 and John Hidalgo Moya b.1920
First big commission after winning the City of Westminster competition for the large Pimlico housing estate, later called Churchill Gardens (1946). Other works include Chichester Festival Theatre (1962), large additions to St John's College, Cambridge (1967) etc. F, H & P   Sensitive, glass-panelled cubism. LB
1. SWINDON. Princess Margaret Hospital 1957   P
2. CORSHAM. Beechfield (Bath Academy of Art) Dining Hall
   kitchens 1968-70   P
3. SALISBURY. Cheese Market. Development - Museum (projected
   1974)   P

POWNALL, F.H architect
1. GRIMSTEAD. East Grimstead. Holy Trinity 1857   P

POWNALL & YOUNG architects
1. WEST DEAN. St Mary's Church 1866   P

POWYS, A R.
1. LUDGERSHALL. St James' Church. Restoration work 1926   WRO

PRANGLEY, John architect and builder of Warminster
1. SHERRINGTON. Parsonage 1826   WRO

PRANGLEY, Thomas architect and builder of Heytesbury
Thomas and John (see above) were timber merchants and builders at Heytesbury in 1830. ETD
1. NORTON BAVANT. Parsonage 1802   WRO
2. TILSHEAD. Parsonage (with John Prangley) 1818   WRO

PREEDY, Frederick architect of London
1. WINSLEY. National School 1868   WRO

PRITCHARD, John architect of Llandaff, Glam.
Diocesan architect.
1. LACOCK. School. Left side extension 1864   P   Work 1869 WRO

PRIVETT, William master mason of Chilmark
In 1768-9 Privett made a design (not accepted) and oversaw the building of Pinford Bridge at Sherborne Castle, Dorset. AS
1. AMESBURY. Amesbury Abbey. Work on W wing 1756-61 (using Chilmark stone)
   RCHM
2. WILTON. Wilton House. Estimate for steps from Venetian window to garden
   1757. Work not carried out.   RCHM

PROVIS, John architect of Chippenham
1. CHRISTIAN MALFORD. Parsonage 1816   WRO

PUGIN, Augustus Welby Northmore designer & architect 1812-1852
Son of emigre French artist. LB Promoted Gothic style. See his "The true principles of pointed or Christian architecture" (1841). Best work: Cheadle Church, Staffs (1841-6); Nottingham R.C. Cathedral (1842-4); St Augustine, Ramsgate (1846-51).   F. H & P

1. ALDERBURY. St Marie's Grange 1835 (built by Pugin for himself after he married, his first new building GW). Contains stained glass designed by him. P Enlargements, possibly by Pugin 1841. (He left the house in 1837 and sold it in 1841. GW)　　　　P
2. CLARENDON PARK. Clarendon House. Gate Lodge. 1837　　　　P
3. SALISBURY. Exeter Street, St Osmond 1847-8　　　　P
4. SALISBURY. New Canal. John Hall's house - restoration 1834 with F.R.Fisher (Fred Bath built a new front in 1881, and the house itself became the entrance of a cinema in 1931). VCH　　　RCHM
5. BISHOPSTONE S. St John Baptist. S transept S window designed by Pugin. "Undeniably uninspired" P   Also designed Gothic tomb 1844 P

PYM, Roland
1. LUDGERSHALL. Biddesden Farm, internal decorations 1930s VCH
2. LUDGERSHALL. Biddesden House. Trompe l'oeil scenes painted in several windows (with D. Carrington) 1930s　　　　VCH

RALPH, (or RALPHS or RALFS) John architect and builder of Warminster
Died between 16 and 23 January 1837.　　　　RG
1. WYLYE. Fisherton Delamere. Parsonage 1833　　　　WRO
2. WARMINSTER. Workhouse, builder to design of Kempthorne 1836-7 RG

RANDELL, Major Albert Joseph architect 1864 - ? of Beech Cottage, and Exchange Place, Devizes
Born Devizes. Entered office of his father John Ashley Randell (qv) who practiced at Exchange Place, Devizes and took it over on his death (Kelly's 1880-1907); 4 St John St (Kelly's 1911). Albert J. Randall listed at Market Place (Kelly's 1915-1927 at least). Many commissions for domestic architecture in Devizes and neighbourhood.　　　　D
1. LITTLE BEDWYN. Kelston House for Gauntlett family (Rendell builder) 1909 R

RANDELL (or RANDALL), James and John builders of Devizes
W. Rendell (qv) did work for them in 1848. R   A James Randell was a grocer, tallow chandler and coal merchant in the Brittox in 1822 and at Maryport Street and Maryport Wharf in 1830 and 1842. ETD

RANDELL, John Ashley architect of Devizes -1898
Took active part in public life of Devizes. D   Office at Exchange Place, Station Road in 1882 (Deacon's Gazeteer).
1. URCHFONT. Parsonage 1859　　　　WRO
2. DEVIZES. Castle. Additions for R.V.Leach 1860-80　　　P
3. BISHOPS CANNINGS. Parsonage 1869　　　　WRO
4. DEVIZES. Northgate Street. Congregational Chapel. Improvements 1876 VCH
5. DEVIZES. Wadworth's brewery 1885　　　　VCH
6. DEVIZES. 29 Market Place. Offices for E.& W.Anstie Ltd 1894 (builder Ash).　　　　WBR   VCH

RAWLINS, builder?
1. SEMINGTON. Whaddon House estate. Work and stone provided 1681 LP

RAWLINS, William builder
1. STEEPLE ASHTON. The Green. Lock-up 1773　　　　WRO

- 78 -

RAYMOND,
1. DEVIZES. St Joseph's Walk. Church of Immaculate Conception. Chancel 1909 (with A.J.Scoles) P

RAYSON, Thomas architect
1. OAKSEY. Woodfolds Farm, altered, restored and enlarged 1938-41 VCH

READ, William Henry architect and surveyor
Practice at 31 Wood Street, Swindon (Kelly's 1875-1899). Worked with W.R.Osborne (qv). Offices at Corn Exchange (1882 advert. in Deacon's Gazeteer).
1. SWINDON. Station Road. Dairy, now Southern Laundry Co. 1876-7 (alterations 1891) PW
2. SWINDON. Cricklade Street. Anderson Hospital (almshouses) 1877 P
3. SWINDON. 33 Wood Street 1883 JO
4. SWINDON. Regent Street. Baptist Tabernacle 1886 (perhaps with R.J. Beswick). Now dismantled P
5. SWINDON. 1 Commercial Road (alts. 1899 for Swindon Permanent Building Society) JO
6. SWINDON. Rodbourne Cheney. Primitive Methodist Church JO
7. SWINDON. Victoria Hospital (with W.R.Osborne) JO
8. SWINDON. Even Swindon Hotel (with W.R.Osborne) JO
9. SWINDON. County Ground Hotel (with W.R.Osborne) JO
10. SWINDON. Devizes Road. Premises for Messrs Chandler JO

REASON, William builder of Marlborough
1. CHISLEDON. Parsonage 1821 WRO

REED and MALLIK builders of Salisbury
1. WILTON. Olivier Road, Reema houses 1949 HHW

REEVE, J.A (or J.R DOE**)** architect of London
1. SALISBURY. St Mark's Infants School 1889, possibly by Reeve DOE
2. RAMSBURY. Holy Cross Church. Restoration work 1890 WRO
3. BAYDON. St Nicholas Church. Restoration 1892 VCH
4. SALISBURY. St Mark's Church 1892-4 (consecrated 1899, but not completed till 1915) VCH
5. WARMINSTER. Church Street, St Boniface College. Addition to the 1796 house 1910 P

RENDELL, F and SONS builders of Devizes
Firm established 1847. Based at 29 St John's Street 1847-1903, 11-12 St John's Street 1903-19/1 and adjoining 2-3 St John's Alley and 13-14 St John's Street 1924/5-1971. Use of Elm Tree Yard from 1904.
William Rendell, bellhanger, gas fitter, whitesmith and locksmith was born at Barnstaple, Devon 1817 and died 1884. Came to Devizes 1842, at end of 1847 moved business from 17 High Street to 29 St John's Street where he had fitters' and blacksmiths' workshops. Sons; Frank J. Rendell born 1852, died 1904 and John Rendell born 1854, died 1927 (based at Northgate Street). After William's death firm called F. and J. Rendell until 1888, then F. Rendell. Frank was a trained smith, also carried out many water and drainage installations at country houses and farm estates. As building firm worked for architects Ponting and Adye. In 1904 sons William James (1878-1952) and Fred (who died 2 months later) took over. William in partnership

with his mother 1906, other brothers in firm. Firm worked especially with architects Brakspear and Ponting in 1920s and 30s. Became Ltd. Co. 1924. After the 1939-45 war firm was involved in returning temporary airfields to agricultural use and restoring requisitioned properties for handing back to owners. 29 St John's Street sold 1952. Joined Farrow Group 1971 which was sold to Y.J.Lovell Ltd c1976-9. Records at WRO.   R

1. DEVIZES. 41 Long Street, wrought hinges and locks to front door 1873 R
2. ROUNDWAY. Roundway House, gas installation 1877   R
3. DEVIZES. Long Street, St John's Church Parish Room, railings, late 19C R
4. DEVIZES. Estcourt Hill footpath, railings by railway foot bridge, late 19C   R
5. DEVIZES. Castle Grounds. Heating system to conservatory for Mrs Brown 1880 R
6. UPAVON. Church, hinges on doors and bell tower, late 19C   R
7. POTTERNE. Capt. Sterne's house, heating system for conservatory 1882, well pump 1883
8. AVEBURY. Beckhampton Stables, extensive work 1898-1906 for Sam Darling, including well 1898. Stables restored for Murless family after wartime occupation c1945   R
9. DEVIZES. 11 (rebuilt) and 12 (restored) St John's Street, first major building job, for own firm 1903-4   R
10. DEVIZES. Town Hall, refurbished and redecorated (architect Ponting) 1903   R
11. HAM. Ham Manor, extensive alterations and repairs, installation of electric light 1909   R
12. LITTLE BEDWYN. Manor, restoration work early 20thC   R
13. LITTLE BEDWYN. Kelston House (first house building contract) for Gauntlett family (A.J.Randell architect) 1909-10   R
14. COULSTON. Baynton House, restoration work early 20thC   R
15. DEVIZES. St John's Church, restoration work early 20thC   R
16. BROMHAM. Rowdeford House, electric light installed early 20thC   R
17. DURRINGTON. Durrington Manor, servants' quarters for Capt. Borwick c1910 R
18. BERWICK BASSETT. Church, new headstocks for bells and ash wheels for ropes 1912   R
19. DEVIZES. Potterne Road, Cambrai and Stavordale, semi-detached pair (W. Rendell moved to Cambrai) 1914-15   R
20. DEVIZES. Caen Hill Brickyard. Foreman's house 1919-20   R
21. DEVIZES. Market Place, Lloyd's Bank refitted 1919-20   R
22. MELKSHAM. Lloyd's Bank rebuilt 1922   R
23. MELKSHAM. King's Arms Hotel, alterations 1922   R
24. WEST LAVINGTON. Dauntsey's School, Headmaster's House mid 1920s. Rendell were main contractor for school late 1920s-early 1930s, built Biological Lab. 1929, new Dining Hall and Farmer Hall 1932. Lavington Manor restored for the school. Extensions 1952. Classroom block 1954. Science block 1956. Lecture Room 1962. Design and build extension to Jeanne House 1978. Gym converted to classrooms 1979. New Sports Hall and Master's House 1980   R
25. DEVIZES. St Mary's Church, restoration of nave roof etc 1924 (architect Brakspear)   R
26. DEVIZES. Potterne Road, Lymington, reusing stonework from Lloyd's Bank, Lymington, completed 1926 and W.Rendell moved there.   R
27. MILTON LILBOURNE. Church, rebuilding of leaning arches and pillars 1920s R

28. DEVIZES. Wick, Wimereux, for Ernest Rendell, late 1920s R
29. DEVIZES. Wick. Row of houses of different designs, inter-war years. 84 houses for MOD c1966 R
30. DEVIZES. Southbroom House, conversion to school for WCC 1926 R
31. DEVIZES. New R.C. School 1929 R
32. DEVIZES. Caen Hill brickyard. Display building 1930 R
33. SAVERNAKE. Savernake Hospital, Nurses' Home 1931 R
34. SALISBURY. Bemerton. Methodist Chapel 1932 R
35. MARLBOROUGH. Marlborough College. Science Block 1933 (architect W.G.Newton), one of first exposed aggregate reinforced concrete buildings in England. Extensions to Priory, Preshute and Upcot House for the school. Extensions 1951. New kitchen staff quarters and groundsman's cottage 1959. Norwood Hall 1960. Art Room and Cloisters 1962. Littlefield House rebuilt 1963. Memorial Hall, entrance and staircase 1964. Summerfield House, extension 1966. Gym and squash courts 1967. Science block alterations 1970. Pair of staff houses 1971. Preshute House modernised 1972 and Barton Hill House extended 1973. R
36. DEVIZES. Assize Courts, restoration works 1934 R
37. LITTLE BEDWYN. Farmer Homes for retired agricultural workers 1936 and 1961 R
38. ROUNDWAY. Roundway Hospital. St Luke's Chapel 1937 WT
39. WILTON. Council housing 1930s R
40. DEVIZES. Edward Road, semi-detached houses 1930s R
41. DEVIZES. Moonrakers' Inn for Wadworth's Brewery 1930s R
42. MALMESBURY. Holloway R.C. School 1930s R
43. DEVIZES. Caen Hill, housing for Devizes Town Council 1930s R
44. LONGBRIDGE DEVERILL. House 1930s? R
45. ROWDE. Cross Keys Inn 1930s R
46. CHIPPENHAM. Sir Audley Arms Inn 1930s R
47. MELKSHAM. Hospital 1937 R
48. STAPLEFORD. Druid's Lodge racing stables, reconstructed pre-1939 R
49. TROWBRIDGE. Nelson Haden School. Part finished 1939 when war started R
50. BISHOP'S CANNINGS. Horton. Pill box by canal 1939-45 period R
51. STEEPLE ASHTON/KEEVIL. RAF Station 1941 (sub-contractor to another builder) R
52. UPAVON. RAF Station, work 1941 R
53. DEVIZES. Flax factory for MOW 1939-45 period R
54. ERLESTOKE. Army camp 1939-45 period R
55. WESTBURY. Victoria College converted to convalescent home 1939-45 R
56. BULFORD. New Reception Hospital 1939-45 R
57. ROUNDWAY. Prince Maurice and Waller camps converted to American Hospital 1939-45 R
58. ROUNDWAY. Horton Road, storage depot for MOW, 30 huts with services 1939-45 R
59. COMPTON BASSETT. RAF camp 1943-6 R
60. LYNEHAM. RAF camp 1946, or earlier, onwards R
61. BROMHAM. Houses 1947 R
62. HULLAVINGTON. RAF station housing 1952 R
63. WROUGHTON. Houses at RAF station 1952. New Officers' Mess. 30 housing units for USAF R
64. CHISELDON. Burderop Park. Nurses' Quarters R
65. UPAVON. RAF station, 101 quarters 1954. Pews for RAF Upavon church after 1945 R
66. MELKSHAM WITHOUT, COLERNE. RAF stations, housing after 1939-45 war R

67. WILCOT. Oare, Flint House, restoration 1940s or 50s   R
68. DEVIZES. Longcroft Road estate 1953 with Cavalier Hotel for Wadworth 1961   R
69. DEVIZES. Nursteed. New Grounds Farm development, sold to MOD, early 1950s
70. DEVIZES. Elm Tree Gardens, mid 1950s   R
71. DEVIZES. Pans Lane development, mid 1950s   R
72. CORSHAM. School 1950s   R
73. MARLBOROUGH. School 1950s   R
74. DEVIZES. Jackson's Close   R
75. MELKSHAM. Shurnhold Secondary Modern School, now George Ward School 1950-1952   R
76. MELKSHAM. Council houses for RDC c1953   R
77. UPAVON. Primary School   R
78. TROWBRIDGE. Girls' Hostel for WCC late 1950s   R
79. STANTON ST QUINTIN. Primary School 1950s?   R
80. CHIPPENHAM. Hardenhuish Lane. Girls' Secondary Modern School, open 1956   R
81. FROXFIELD. Somerset Hospital, alterations and additions 1957
82. DEVIZES. 15 St John's Street and 17 High Street, joinery 1958   R
83. MARLBOROUGH. Church of St Thomas More, rebuilt 1959   R
84. GREAT BEDWYN. The White House, restored 1959   R
85. DURRINGTON. Larkhill Camp, annexe to Medical Centre late 1950s or early 1960s   R
86. DEVIZES. Police Hostel for WCC late 1950s   R
87. DEVIZES. London Road, new Police HQ 1961   R
88. DEVIZES. Heathcote House and Southbroom. School for WCC early 1960s   R
89. MARLBOROUGH. Nurses' Home early 1960s   R
90. ERLESTOKE. 18 prison officers' houses at former army camp early 1960s. Two houses 1964. Two accommodation blocks 1981   R
91. MELKSHAM. Barclay's Bank, modernised 1960s   R
92. CHIPPENHAM. Kingfisher Pub. for Wadworth 1964   R
93. CORSHAM. Flats (architect Wyvern Design Group) 1965   R
94. TROWBRIDGE. Post Office 1965   R
95. TIDWORTH. 38 Officers' Married Quarters 1966   R
96. MARKET LAVINGTON. Estate 1968   R
97. ALDBOURNE. Upham House, doors   R
98. DEVIZES. Downlands estate, 'Farrow Method' housing 1971   R
99. SWINDON. Eldene estate, Partnership Housing with Swindon Borough C. 1971   R
100. GRAFTON. Wilton Windmill, repair of structure 1971 (architect Bowden) R
101. SWINDON. Steel-framed office and warehouse for Kennedy c1972   R
102. SALISBURY. Godolphin Girls School. Hall of residence c1972. Library extension to make 6th Form centre 1977   R
103. SWINDON. Temple Street, 11 storey office block for Aspen Developments (Rendell's tallest building) 1973   R
104. SWINDON. Liden development with Borough of Thamesdown, early 1970s. Subsequent similar developments at Liden, Merlin Way and Kingshill. Later Toothill (52 houses, then 82 then 85) pre-1977. Freshbrook 'urban village' 1979-81   R
105. SWINDON. High Street and Victoria Road. Offices for Sun Life Assurance 1970s   R
106. MARLBOROUGH. Salisbury Road, development of 60 houses 1970s   R
107. SWINDON. Okus estate, fourth phase (11 houses, 1 bungalow) 1970s   R

108. DEVIZES. Rotherstone, third phase for local authority 1970s   R
109. SWINDON. Wrenswood, 36 flats for Kennet Housing Soc. 1970s?   R
110. WOOTTON BASSETT. Conversion of Sec. Mod. School to Comprehensive and additional buildings for WCC 1970s   R
111. SALISBURY. Portman Building Soc. building, repairs   R
112. MELKSHAM WITHOUT. Bower Hill estate for MBRDC 1974 and for WWDC later R
113. EVERLEIGH. Church, repairs 1975   R
114. KEEVIL. Church, repairs c1975   R
115. DEVIZES. New Park Street, Brownston House, leadwork restored for KDC 1976   R
116. SALISBURY. The Close, Mompesson House, repairs and redecoration for NT   R
117. URCHFONT. Energy-saving houses c1978   R
118. WESTBURY. Eden Vale, Partnership housing with WWDC at former brickworks clay pit, 44 houses then 30 more units 1979   R
119. ROWDE. 30 houses for KDC 1979   R
120. BROMHAM. Greystones private estate 1980   R
121. MELKSHAM. Housing for WWDC 1980-1   R
122. AMESBURY. Housing for SDC 1980-1 (71 houses)   R
123. ALDBOURNE. Housing for KDC 1980-1   R
124. CHIPPENHAM. Housing for NWDC 1980-1   R
125. WESTBURY. Bitham Park estate. 300 houses built on 30 acres 1981   R
126. SWINDON. Office block for James Davies (Timber) Ltd c1981   R
127. MARLBOROUGH. Swimming Pool for KDC post-1980   R
128. TROWBRIDGE. Weavers Drive, retirement homes 1982   R
129. HULLAVINGTON. Cricket Pavillion for RAF 1983   R

RENNIE, John engineer 1761-1821
Born in Scotland, after training set up in London. Famous for Kennet & Avon Canal in Wiltshire, but also interested in fen drainage, harbours and docks, lighthouses and bridges (Waterloo Bridge - 1810, and other London bridges)   CH   F H &P
1. WESTWOOD. Avoncliff. Aqueduct before 1804   P
2. WINSLEY. Dundas Aqueduct c1805   P
3. WILCOT. Ladies Bridge 1808   P

REVETT, Nicholas painter & architect 1720-1804
Society of Dilettanti. Private means, so did not practise extensively.
1. DOWNTON. Standlynch. Trafalgar House. The porch and the lobby in the N pavilion. After 1766   P   C

REYNOLDS, Evan of Trowbridge
1. TROWBRIDGE. Town Bridge 1777   P

REYSSCHOOT, Pieter Jan van painter
1. WILTON. Wilton House. Painting of N side staircase 18thC   RCHM

RICHMAN, George & CO masons
1. SEMINGTON. Whaddon House estate. Work 1763

RIGAUD, J.F painter
1. CALNE WITHOUT. Bowood House. Frescoes in grisaille for Robert Adam c1769-1773   WAM 41

RIGBY, J & C builders of London
Contractor already employed before 1841 on GWR works and stations.
1. SWINDON. Railway village comprising a number of streets with terraces and larger houses for senior staff, about 300 houses in all (architect Brunel) 1841-1870   DOE

ROBBINS, S architect and **BIGGS, Ben**
1. STANTON ST BERNARD. Honey Street, Barge Inn 1856   WBR

ROBERTS, David H.P architect
1. TROWBRIDGE. College of Further Education 1957-9   P
2. MARLBOROUGH. College, Norwood Hall (dining hall) 1959-61 P   1961-2 VCH

ROBERTS, Edward architect of London
1. SWINDON. Mechanics' Institute 1853-4 (but largely rebuilt 1892). Gothic Revival style   P
2. URCHFONT. St Michael's Church. Restoration work 1863   WRO

ROBERTS, Ernest S architect of Birmingham
1. DEVIZES. New Park Street, Regal Cinema 1939. Now demolished   WBR

ROBERTSON, E architect of Swindon
1. SWINDON. Market Square, Corn Exchange (with S Sage) 1852-4

ROBERTSON, William builder of Bristol
1. CALNE. Workhouse 1848 (architect Allom)   RG

ROBINSON & KAY architects
1. CALNE. Church Street. Building opposite the old Capital and Counties Bank 1953-4   P

ROBSON, John architect
1. RAMSBURY. Littlecote House. Alterations to the Library C19   VCH

ROE, William Henry of Southampton, Hants.
1. SALISBURY. Brown Street, Baptist Chapel 1829   RCHM

ROGERS AND BOOTH builders of Gosport, Hants.
1. SUTTON VENY. St John's Church 1866-8   JO

ROGERS, Edward surveyor
1. MARLBOROUGH. St Peter's Parsonage 1813   WRO

ROGERS, H.S architect of Oxford
1. SHERRINGTON. Sts Cosmos and Damian. Restoration work 1930   WRO

ROLFE & PETER architects of Bath
1. WESTWOOD. St Mary's Church. Restoration work 1934,6   WRO
2. BRADFORD-ON-AVON. Budbury Farm. Rebuilt 1925 (drawings British Architectural Library, London)   JO

ROMAIN, J builder of Devizes
Moved to Sussex Wharf on the canal c1898 (Gillman's Directory).   VCH

ROSE, John plasterer
Worked for Robert Adam.
1. CALNE WITHOUT. Bowood House. Ceilings and wall panels c1760-1765
   WAM 41

ROSEBLADE, John builder of Latton
1. LATTON. St John the Baptist's Church. Alterations to chancel etc 1861 DH

ROSETTI, Dante Gabriel stained glass designer and painter 1828-82
1. MALMESBURY WITHOUT. Rodbourne. Holy Rood Church. E window, designed with Ford Madox Brown and made by Morris & Co 1865   VCH

ROSGRAVE, painter and carver of Salisbury
Humphrey Beckham (qv) was apprenticed to him c1605. VC

ROSS, W.A architect
Chief architect to the War Office.
1. DURRINGTON. Larkhill camp. St Alban's Church 1937-8   VCH   P

RUDMAN, Walter architect of Chippenham   -1939
Served in First World War. Acquired practice of Holloway (qv) & Fogg c1921. Practice at his home, Zealy's House, 53 St Mary Street, Chippenham, (Kelly's 1923). Moved to top floor of Jubilee Institute, 32 Market Place in 1925 and practice there changing name to Edwards and Webster in 1948 until removal when Wyvern Design Group formed in 1952. P.W.Edwards (qv) had become partner in 1938. PE
1. CHIPPENHAM. Greenway Lane. The Priory (for himself) 1910 JO
2. CHIPPENHAM. Greenway Park, Greystones c1910   JO
3. CALNE. Quemerford, The Croft (formerly Moon Croft). Extension c1918
   NMR?
4. BIDDESTONE. The Close. Work 1923   PE
5. ENFORD. East Chisenbury Priory. Work 1923   PE
6. CHIPPENHAM. Cottage Hospital. Alterations c1925-39 (plans) WBR
7. CHERHILL. St James' Church. Restoration work 1926   WRO
8. LACOCK. Cantax Hill. Dyer's Leaze 1930   JO
9. LYNEHAM. St Michael's Church. Restoration work 1936   WRO
10. CALNE. St Mary's School 1936   P
11. CALNE WITHOUT. Derry Hill. Christ Church. Restoration work 1937   WRO
12. CHIPPENHAM. Lowden Hill. Lowden Manor. Alterations (plan)

RUNDELL, Henry mill carpenter of Wingfield
Lease of cottage at Ladywell, Newtown, Bradford-on-Avon 1686. MP

RUSHTON, Henry carpenter of Turleigh, Winsley
1. BRADFORD-ON-AVON. 10 Belcombe Place. Bought empty plot in 1825 and probably built house (by 1828) with T.Woodman (qv).   PMS

RUTHERFORD, Thomas architect and builder of Calne
Stone mason at the Quarry, Calne in 1842   ETD
1. CHERHILL. Yatesbury. Parsonage 1841   WRO
2. CALNE. Old workhouse. Alterations (Salway architect). Rutherford said to be of Chippenham at this time.   RG

RYDER, Captain Richard architect and carpenter
Work at Cranborne Manor House, Dorset 1647-50 and possibly at St Giles's House, Wimborne, Dorset in 1651.
1. WILTON. Wilton House. Master carpenter for Webb's refitting of house after 1647 fire   RCHM

SABBATINI, Lorenzo painter
1. WILTON. Wilton House. Ceiling painting c1650   RCHM

SAGE, Sampson architect of Swindon
1. SWINDON. Rodbourne Cheney. St Mary's Church. Restored and enlarged 1848   JB P
2. SWINDON. Market Square. Town Hall 1852 (with E. Robertson). Originally built as a corn exchange, but never used as such. Found inadequate and rebuilt or extended 1865-6   DOE   VCH 'Remarkably purely Grecian for its date' P
3. BROAD TOWN. National School 1858   WRO

ST AUBYN, J.P architect
1. DILTON MARSH. Chalcot House. Alterations and extensions for Phipps family 1872 P   Probably also stable block   DOE

ST PANCRAS IRON WORKS CO ironfounders
1. CHIPPENHAM WITHOUT. Sheldon Manor. Stable fittings for former barn c1911 SP

SAINSBURY, James builder of West Lavington
1. EASTERTON. St Barnabas Church (to design of Mr Christian) 1875 SJ

SALMON, Charles stone mason of Devizes
Gillman's directory 1848.
1. DEVIZES. The Green, Victoria Cottage. 1850s or 60s? Elaborate small house with stone statues etc. to advertise his trade.   WBR

SALVIN, Anthony architect 1799-1880 (or 81)
One of the most prolific of Victorian architects. GW An authority on the restoration and improvement of castles (inc. Tower of London, Windsor, Caernarvon, Durham, Warwick, etc). But a domestic architect as well, especially country houses. Specialized in Elizabethan and Jacobean styles.
LB   F, H & P
1. STRATTON ST MARGARET. Church? Restoration work 1845   WRO
2. ODSTOCK. Longford Castle. Rebuilt facade c1870, and C17 office outbuilding restored (and rebuilt?). Restored again 1973. P

SALWAY (or SOLWAY), George Sargent architect and surveyor of Chippenham
Carpenter and builder at River Street in 1842. ETD   G. Salway, builder, of Chippenham repaired Dauntsey Bridge 1827. PL
1. CALNE. Old workhouse, plans and specs. for alteration   RG
2. HILMARTON. Parsonage and coachhouse 1841   WRO
3. NETTLETON. Nettleton and Burton School and house 1848, addition 1849. Dated 1850 (The Builder 1849 p449)   JO WRO

SARJEANT, Anthony architect of Wimborne, Dorset
1. STOCKTON. Rectory 1790-2   VCH

SCHAFFLIN, Robert plasterer
1. SALISBURY. General Infirmary 1768				RCHM

SCHMIDT, architect
1. BROUGHTON GIFFORD. Manor House. Alterations c1908		WBR

SCOLES, A.J architect
An earlier architect Joseph John Scoles did work at Bath in 1844 and was a leading Roman Catholic architect, travelling abroad in the early 1820s. NJ
1. TROWBRIDGE. Wingfield Road, St John the Baptist (R C) 1875 P
2. DEVIZES. Immaculate Conception (R C). New chancel 1909 (with Raymond) P

SCOTT, Sir George Gilbert architect 1811-1878 of London
Started with workhouses (with W.B.Moffatt) and the same partnership went on to churches, the first being St Giles' Camberwell. Best work included Albert Memorial (1864) and St Pancras Station Hotel (1865). Scott was a Gothicist, a disciple of Pugin, acceptable to the Cambridge Camden Group. Became Surveyor of Westminster Abbey 1849. Also active as a secular architect, again in Gothic style, but was forced by Palmerston to build the Government offices in Whitehall (1861) in Renaissance style. Knighted 1872. NJ, SPC, LB, F H &P "A safe copier of other people's precedents." JB (See also Scott and Moffatt)
1. ZEALS. St Martin 1842-4					P
2. SWINDON. Cricklade Street, Christ Church 1851		P
3. CHIPPENHAM. St Paul's Church 1853-61 (consecrated 1855) SPC P
4. BRATTON. Vicarage 1863 (Kelly 1867)				JO
5. SALISBURY. Cathedral. Restoration of West front, Lady Chapel, transepts and choir 1863-67						VCH
6. SALISBURY. 60 The Close (North Canonry). Restoration and addition. P
7. SALISBURY. St Edmund. Chancel and South chapel 1865-7	P
8. REDLYNCH. Hamptworth. Common. National School 1867		VCH
9. CHIPPENHAM. Malmesbury Road, St Paul's Rectory (attributed) 1868 SPC
10. BRATTON. Church of St James. Rebuilding of chancel (in association with T.H.Wyatt) WRO
11. SAVERNAKE. Savernake Cottage Hospital 1871-2		P

SCOTT George Gilbert Jnr architect 1839-97
Son of Sir George Gilbert Scott.
1. GREAT BEDWYN. Glebe House (originally Vicarage)		JO
2. MALMESBURY WITHOUT. Burton Hill House, int.alts. c1880	JO

SCOTT, Sir Giles Gilbert architect 1880 1960
Grandson of Sir G.G.Scott. Best known for Liverpool Cathedral. CH Designed red telephone kiosks (now listed) and electricity pylons (1930s).
1. STEEPLE LANGFORD. All Saints. Lych Gate 1902			VCH

SCOTT, John Oldrid architect 1842-1913
Son of Sir George Gilbert Scott. Gothicist like his father, best known for Catholic Church of Norwich, 1884-1910.   F,H & P
1. BRADFORD-ON-AVON. Christ Church, Bearfield. Chancel 1878 P

SCOTT, G.G & MOFFATT, W.B architects
The Scott became Sir George Gilbert Scott; Moffatt, W.B. (See also separate entries.)

1. AMESBURY. Workhouse 1837   RG
2. MERE. Workhouse 1840   RG
3. TEFFONT. Teffont Evias. Rectory 1842 P. Probably by partner Moffatt 1841   JO
4. SWALLOWCLIFFE. St Peter 1842-3   P
5. SWINDON. St Mark's Church 1843-5   P
6. SWINDON. Church Place, St Mark's Vicarage c1845   PW
7. SWINDON. GWR National School- Infants 1856   WRO PW

SCOTT, William carpenter
1. DAUNTSEY. Rectory 1829-33   DR

SEARCHFIELD, Thomas G builder of Heytesbury
1. HEYTESBURY. National School and house 1860   WRO

SEDDON, J.P architect 1827-1906
Gothic Revivalist (though Voysey was one of his pupils).   G
1. UPAVON. St Mary. Nave restoration 1875 (chancel restoration was by Wyatt, same year)   P
2. CHIRTON. Parsonage 1878   WRO
3. STANTON ST BERNARD. Vicarage altered or rebuilt 1878   JO
4. SWINDON. Gorse Hill, Cricklade Road, St Barnabas 1885   P

SHACKLE, G.H architect
Practice at 7 Alexandra Terrace, Marlborough (Kelly's 1907-1915)

SHAW, John architect
1. MALMESBURY WITHOUT. Cowbridge House, with Italianate S garden front 1853   VCH

SHAW, Richard Norman architect 1831-1912
Pupil of William Burn, became a very successful country house architect. Academy Gold Medal 1854, later worked for Street. Started in practice with Eden Nesfield - Gothicist at first before mature style based on mid-C17 and William & Mary. Shaw designed the earlist garden suburb - Bedford Park, Turnham Green.   F,H & P
1. MARLBOROUGH. College, Upcot House 1855-6   P

SHINGLER RISDON ASSOCIATES i.e. RISDON, Frank Heriot architect of Soho Square, London.
1. SWINDON. Town centre shopping area 1960 (consultant architect Frederick Gibberd). (Shingler & Risdon, architects to the property developers; acted as assistants to the borough engineer and surveyor as planners, and to the corporation's consultant architect, Gibberd.)   VCH

SHIPWAY family builders of Green Street, Avebury. 18thC - 20thC
Also stonemasons. Last was James Shipway. Worked with Titcombe and Paradise families.   SA
1. AVEBURY. Swindon Road, houses   SA
2. AVEBURY. St James's Church, repairs, and lych-gate 1899   SA

SHURMER, T.M architect of Andover, Hants. Died 1872
1. ALLINGTON. Boscombe, Vicarage, work 1836   WRO JO

SILCOCK, Thomas Ball architect of Bath
Born 1854, St Margaret's Place, Bradford-on-Avon. Office in Milsom Street, Bath, and lived on Widcombe Hill. Elected MP for Wells, Somerset, in 1906. RM (See also SILCOCK & REAY)
1. BRADFORD-ON-AVON. Junction Road, Fitzmaurice School 1897   KR and MS
2. TROWBRIDGE. Newtown Junior School 1900   P
3. BRADFORD-ON-AVON. Frome Road, Hillside House (?)   KR and MS

SILCOCK & REAY architects
(See also SILCOCK, T.B.)
1. SWINDON. Sanford Street, Congregational (United Reformed) Church 1894
   P JO
2. CHIPPENHAM. Pew Hill House 1895 (now part of Westinghouse)P
3. SWINDON. Victoria Road, Old Technical College 1897   P
4. CHIPPENHAM. Ivy Lane School c1900   WRO
5. WINSLEY. Winsley Manor 1902 (now Dorothy House)   JO
6. WINSLEY. Chest Hospital 1903-4   P
7. MARLBOROUGH. St Mary's & St Peter's Junior School 1905 (formerly the
   Grammar School)   P

Marlborough Grammar School soon after construction (postcard)

SILLEY, George M architect
1. CHIPPENHAM. National Westminster Bank 1876   P
2. DURNFORD. Great Durnford. Manor House, N service wing 1913 DOE VCH

SILVERTHORN, Anthony carpenter of Whaddon Elm, Hilperton
Born c1660 at East Town, West Ashton, married c1679, died 1728?
(Family History of the Silverthorns)

SILVERTHORN, John carpenter
1. SEMINGTON. Whaddon House and estate. Work 1679-80 and 1682 LP

SINGLETON, George of Donhead St Andrew
1. EBBESBOURNE WAKE. Fifield Bavant. Parsonage 1845   WRO

SLATER, William architect of Regent Street, London
1. STEEPLE LANGFORD. All Saints Church. Chancel rebuilt c1857 VCH
2. STEEPLE LANGFORD. School and house 1859   WRO
3. DEVIZES. St John's Church. West front 1861-2   P
4. SALISBURY. St Thomas. School 1863   WRO
5. CALNE. St Mary's Church. South porch & transept rebuilding 1864 P
6. CHERHILL. Parsonage 1864   WRO
7. BROMHAM. St Nicholas Church. Chancel rebuilding 1876   P
8. LACOCK. Cantax Hill, Old Rectory (possibly by Slater)   JO

SLATER, W & CARPENTER, R.H architects of London
1. BROMHAM. St Nicholas Church. Restoration work 1865   WRO
2. DEVIZES. St Peter's Church 1866-7   P
3. CALNE WITHOUT. Quemerford. Trinity Schools 1868   WRO
4. HANNINGTON. St John the Baptist 1869-71   P

SLEAT, William architect and surveyor of Salisbury
At Exeter Street in 1822, 1830. Architect and surveyor of same name at Market Place 1842. ETD   Made plan of Tailors' Hall, Milford Street in 1823. RCHM (Salisbury vol 1)
1. STEEPLE LANGFORD. Little Langford Parsonage 1827   WRO
2. SHREWTON. Parsonage 1828   WRO

SMEATON, John civil engineer
1. WILTON. Wilton House. Baluster Bridge 1777   RCHM

SMITH Cyril H architect of Calne (1879-?)
Articled H.T.Holloway of Chippenham (Who's Who in Architecture 1926)
1. CALNE. Secondary School   JO
2. CALNE. Free Library   JO
3. CALNE. The Highlands   JO

SMITH, H Stanley architect
1. SWINDON. Cheney Manor estate. Semiconductors factory 1957-8   P

SMITH, James mason
1. STANTON ST QUINTIN. Upper Stanton Farm, repairs 1836   EG

SMITH, John James architect and surveyor of Swindon
Practice at Eastcott House, New Swindon (Kelly's 1875); in Faringdon Street (Kelly's 1899).
1. SWINDON. Turner Street (builder Turner) 1894   JB
2. SWINDON. Hunt Street, terrace (builder Turner) 1895   JB

SMITH, Thomas carpenter and builder of Highworth
(See also W. Pedley.)
1. STRATTON ST MARGARET. Highworth and Swindon Union old workhouse,
   alterations (T.Angel architect)   RG

SMITH, William architect and builder of Trowbridge
Yard in Church Street on site where Tabernacle later built. ML. Carpenter

at The Halve in 1842 ETD Leading Trowbridge architect and builder in later 19C to whom most of the Victorian Gothic buildings in the town are probably attributable. A man of exceptional talent.   JO
1. TROWBRIDGE. Conigre, Unitarian Chapel 1855 (1857-9 JO)   P
2. TROWBRIDGE. 24 Fore Street, shop for Wm. Beaven, ironmonger, probably late 1864   KR
3. TROWBRIDGE. Wyke House. Rebuilding c1864   MMa
4. TROWBRIDGE. St Thomas 1868-70 "Just a little nightmarish" P Also nearby lodge and school   JO P
5. MELKSHAM WITHOUT. Shaw. New school and house 1871 (builder W.H.Bromley) WRO
6. MELKSHAM WITHOUT. Sandridge and Forest. School and house 1872 WRO
7. TROWBRIDGE. Stallard Street, office building 1878 and adjoining cloth hall for wool sales 1869 at Clark's Studley Mill.   JO
8. TROWBRIDGE. Armoury 1877   JO
9. TROWBRIDGE. The Down. Avon View House 1870s? Demolished c1961   ML
10. STAVERTON. Village School   JO
11. TROWBRIDGE. Emmanuel Baptist Chapel, Church Street Schools added behind 1884 (plaque)   JO
12. TROWBRIDGE. Cemetery, Kingston and Hastings Mausoleum   MMa

SMYTH, Robert carpenter of Salisbury
Freeholder of city in 1607-8. WAM 19

SMYTHSON, Robert mason and architect c1536-1614
The only Elizabethan architect of note. First heard of at Longleat, but his masterpiece, Wollaton Hall, Notts. built in next decade (1580-8). See also Worksop Manor, Hardwick Hall and Burton Agnes. F,H & P
1. HORNINGSHAM. Longleat. Working there from 1568   P
2. TISBURY. Wardour Old Castle. Minor work there from 1576   P

SNAILUM, Walter Wadman architect of Trowbridge -1934
Practice at 5 Church Street (Kelly's 1899-1927 at least).
1. TROWBRIDGE. Church Street, Auction Rooms 1893   JO
2. UPTON LOVELL. Upton Lovell Mills, rebuilt 1899   JO
3. TROWBRIDGE. The Parade, some Usher's brewery property 1919 P
4. SEMINGTON. Melksham Union Workhouse, vagrants' cells 1925-6 RG
5. TROWBRIDGE. Hospital 1927-9 (with A.J.Taylor and A.C.Ford of Bath) VCH
6. TROWBRIDGE. Gloucester Road, Bethesda Chapel 1930   VCH
7. NORTH BRADLEY. New Baptist Chapel   VCH
8. TROWBRIDGE. St James. Restoration work 1934   WRO
9. NORTH BRADLEY. St Nicholas. Restoration work 1934   WRO
10. WINGFIELD. St Mary. Restoration work 1935   WRO
11. HOLT. St Katherine. Restoration work 1936   WRO
12. SOUTH WRAXALL. St James. Restoration work 1936   WRO
13. STAVERTON. St Paul. Restoration work 1938   WRO

SNOW, William plasterer
Worked for Robert Adam.
1. CALNE WITHOUT. Bowood House. Ceilings and wall panels c1760-65   WAM 41

SOANE, Sir John architect 1753-1837
Trained under Dance and Holland, and studied 3 years in Italy, but French influence probably more profound. Appointed Surveyor to the Bank of England 1788. One of three 'attached architects' to Board of Works in 1814. GW "Stripped and abstracted classicism". NJ More "romantic" later, viz Pitzhanger Manor 1800-3, Dulwich College Art Gallery 1811-14, and his own house No 13 Lincoln's Inn Fields 1812-3. Professor of Architecture, Royal Academy, 1806. Knighted 1831. F, H & P
1. TISBURY. Old Wardour Castle. A boudoir   P
2. TISBURY. Wardour Castle. Chapel. Lengthened by Soane in 1789-90 P
3. NETHERAVON. Netheravon House. Additions 1791 (post 1791 - VCH) P
4. CHIPPENHAM. Hardenhuish House. Soane made designs 1829 "and the porch seems to be his".   P

SOLWAY, George (see SALWAY)

SOPPITT, James architect of Tout Hill, Shaftesbury, Dorset
1. TISBURY. Parsonage 1859   WRO
2. TISBURY. Two school houses 1863   WRO
3. TISBURY. Infants School 1872 (Possibly SUTTON MANDEVILLE. Chickgrove, 1872, illustrated in British Architect 3 of 1875)   WRO
4. COMPTON CHAMBERLAYNE. St Michael. Restoration work 1877 WRO

SPENDER, James builder
1. BRADFORD-ON-AVON. Town Hall 1855 (co-builder with J. Long, architect Fuller)   YP

SPICER, William mason
Later Surveyor of the Queen's Works.
1. HORNINGSHAM. Longleat. Working there in the 1550s (?)   P

STANLEY, William Henry architect
Practice at Town Hall, Trowbridge (Kelly's 1889-1923 at least).
1. CORSLEY. St Mary Temple 1902-3   P
2. NORTH BRADLEY. Brokerswood. Plan of site for erection of iron chapel, C of E 1904   WRO
3. TROWBRIDGE. St James. Restoration work 1907   WRO

STANSFIELD, B John architect
Practice at Town Hall Chambers, Bradford-on-Avon (Kelly's 1899); at Turleigh Hall (Kelly's 1915).

STENT, John builder or architect of Warminster (?)
1. WARMINSTER. 24 Boreham Road and Boreham Terrace. Third party to deed 1829. May have designed and built house and adjoining terrace in 1822 WBR
2. WARMINSTER. Workhouse outbuildings c.1836-7   RG

STENT, Sidney architect
Said to be "formerly of Warminster" on 14.7.1894 (Warminster and Westbury Journal, p5). Designed HQ for All Saints Sisters, Capetown diocese, SA at this time. The Bishop there was brother of Mr H.P.Jones of Portway House, Warminster.

STENT, Thomas builder? of Yeovil, Somerset
1. BARFORD ST MARTIN. School 1853　　　　　　　　　　WRO

STENT, William carpenter and builder of George Street, Warminster 1842 ETD

STENT, William Jervis architect of Warminster
Son of John Stent? Non-conformist architect. RG Earliest chapel at
Lymington, Hants. CS Practice at Portway, Warminster (Kelly's 1875, 1880).
1. WARMINSTER. Market Place, Savings Bank (?) 1852　　　JO
2. EBBESBOURNE WAKE. Congregational Chapel 1857　　　RCHM
3. WARMINSTER. High Street. Athenaeum 1858　　　　　　P
4. SALISBURY. Endless Street, Congregational Chapel. Refronted c1860 (now
　　gone)　　　　　　　　　　　　　　　　　　　　　　RCHM
5. WARMINSTER. George Street. Methodist Chapel 1861　　P
6. BROAD CHALKE. Congregational Chapel 1862-3　　　　RCHM
7. CHIPPENHAM. Foghamshire, Temperance Hall 1863　　　JO
8. MERE. St Michael. Restoration work 1865　　　　　　　WRO
9. CALNE. Church Street, Free Church 1867. "Terrible"　　P
10. MALMESBURY. Westport. Congregational Chapel c1867　P
11. MERE. Congregational Chapel 1868 (RCHM says 1852)　P
12. WESTBURY. Prospect Square c1869　　　　　　　　　　JO
13. WESTBURY. Bitham Mills (?) 1869　　　　　　　　　　JO
14. TROWBRIDGE. Newtown, Methodist Chapel 1872　　　　P
15. WESTBURY. Bratton Road, Laverton Institute 1873. Venetian Gothic -
　　"no mean architect"　　　　　　　　　　　　　　　JO
16. BRIXTON DEVERILL. Parsonage 1876　　　　　　　　　WRO
17. HOLT. New Congregational Chapel 1880. Early English style P
18. SALISBURY. Brown Street, Baptist Chapel, rebuilt or reconstructed
　　1881-2　　　　　　　　　　　　　　　　　　　　　CS
19. WESTBURY. Laverton Infants' School 1884　　　　　　JO

Laverton Institute, Westbury c1980? (WBR Collection)

STEVENS, E architect
1. BROMHAM. Spye Park. Main house refronted c1767 by E.Stevens, and was demolished 1868   JO

STEWART, George architect
1. ERLESTOKE. Erlestoke Park 1786-91 (only two wings remain) P

STILLMAN & EASTWICK-FIELD architects
1. MINETY. Minety House (of early C19), farm buildings 1955 G (p112) P
2. MARLBOROUGH. Children's Convalescent Hospital (former workhouse). Additions to Sir G.G.Scott's 1837 building, 1955-7   P

STOCKING, Thomas plasterer of Bristol
1. BRADFORD-ON-AVON. Belcombe Court. Study ceiling attributed to Stocking   DOE
2. CORSHAM. Corsham Court. Armorial Rococo ceiling in former library P

STOCKING, William plasterer
1. BRADFORD-ON-AVON. Belcombe Court. Drawing room ceiling attributed to Wm Stocking.   DoE

STOCKWELL, Edward architect 1874-? of Goddard Avenue, Swindon
In partnership with H.E.Nicholls (qv). Also associated with James Robinson, County Architect of Hants. and his succesor William J.Taylor. Designed various important public buildings in Hants.   D
1. SWINDON. Ferndale Road School (with Nicholls)   D

STRAPP, John architect
Chief engineer of the LSWR
1. SALISBURY. Market House (Corn Exchange). East terminal of the Market House Railway. E. front 1859 RCHM (Converted to a new library 1972-3)   P

STREAT, John architect of Shrivenham, Oxon.
1. CHISELDON. Parsonage 1839   WRO

STREET, George Edmund architect 1824-1881 of London
Pupil of George Gilbert Scott. Started a practice in Oxford 1852; Webb and Morris were among his first assistants. High Church, so appreciated by the Cambridge Camden Group. Moved to London 1855. Church work Continental Gothic rather than English Gothic. Best known secular work - Law Courts, 1866. F,H & P In London - heavy ,serious and worthy buildings. LB
1. UPTON SCUDAMORE. St Mary. Rebuilding 1855   P
2. COLLINGBOURNE DUCIS. St Andrew. Rebuilding 1856   P
3. BAYDON. St Nicholas. Rebuilding of north aisle 1857-8   P
4. CORSHAM. St Bartholomew. Chancel arch   P
5. MILTON LILBOURNE. St Peter. Restoration work 1859   WRO
6. PEWSEY. St John the Baptist. South Chapel 1861   P
7. WOOTTON RIVERS. St Andrew. Rebuilding 1861   P
8. PEWSEY. National School and house 1861 (builder W.H.Penning), enlarged since.   WRO JO
9. WARMINSTER. Boreham, St John the Evangelist 1865   P
10. CHAPMANSLADE. SS Philip & James 1866-7. H.R.Goodhart-Rendel called

       the church "a model of its kind".   JO   P
11. SALISBURY. St Thomas of Canterbury. Chancel restoration 1866-7 P
12. CHAPMANSLADE. School 1866-7   JO
13. YATTON KEYNELL. St Margaret. Restoration 1868   P
14. WOOTTON BASSETT. St Bartholomew & All Saints. Restoration 1869-71 P
15. MARLBOROUGH. College, Littlefield 1870-2   P
16. WARMINSTER. St John's School 1872 (lych-gate 1874 JO) VCH
17. BRITFORD. St Peter. Restoration 1872-3   P WRO
18. MARLBOROUGH. St Mary. Chancel restoration 1873-4   P
19. MELKSHAM. Forest, St Andrew 1876 (Note: P says ADYE)   VCH
20. MELKSHAM. The former vicarage (now called The Grange) remodelling
    1877   P
21. ERLESTOKE. St Saviour 1877-80   P
22. ERLESTOKE. Vicarage 1877-80   P
23. MELKSHAM. Tithe Barn. Conversion to Parochial schoolroom 1878 JO
24. HILMARTON. St Lawrence. Restoration 1879-81   P
25. SALISBURY. Cathedral. North Porch. Restoration 1880-1   P
26. MELKSHAM. St Michael. Chancel restoration 1881   P
27. MARLBOROUGH. College, Museum Block. Designs, but Street died 1881,
    block built 1882-3 by his son A.E.Street and A.W.Blomfield.
    P   VCH

STREET, A.E architect.
Son of G.E.Street.
1. MARLBOROUGH. College. Muesum Block 1882-3 (with A.W.Blomfield to his
   father's designs, see above)   VCH

STRONG, Thomas architect, quarry owner and builder of Box
On 1840 tithe award given as quarry owner. Aged 60 on 1841 census, builder.
Robert Strong, quarry master in 1859 (Kelly) and 1865 (Harrods Directory).
J. Peniston, county surveyor, in correspondence with Mr Strong at Box about
repair of various county bridges 1824, complains Strong slow to carry out
work. Worked stone being supplied for country house work in other counties.
PL
1. SALISBURY. Council House. Entrance hall columns and windows in Grand
   Jury Room 1828   PL
2. DAUNTSEY. Parsonage 1829   WBR   WRO

STRONG, W architect
1. GREAT SOMERFORD. New rectory 1829-33. Tudor style   VCH

STURGESS, Edward architect of Cholderton
1. NEWTON TONEY. School 1857   WRO
2. BULFORD. C.E.School, school house 1874   WRO

STYLE, A.J architect of Thames Ditton, Surrey
1. SEEND. Holy Cross. Chancel rebuilding 1876   P
2. WILCOT. Holy Cross. Mostly rebuilt by Style after a fire in 1876 P
3. EASTERTON. St Barnabas. Restoration work 1905   WRO
4. DAUNTSEY. Church. Faculty of 1879 (carried out?)   JO

SUMSION, Thomas mason of Colerne c1672-1744
One of the last master masons to carry on the medieval tradition of design.

   C Buried Colerne churchyard. Will proved Oct. 1745.   VH
1. SHERSTON. Holy Cross. Tower 1730, Gothic with battlements
   and pinnacles. P   Paid £1.15.0 for his 'draught'.   VH

SURMAN, Robert bricklayer and tiler
1. SALISBURY. General Infirmary 1768   RCHM

SUTTON, Basil architect
1. RAMSBURY. Holy Cross. Restoration work 1929   WRO

SUTTON, John surveyor
1. HORNINGSHAM. Parsonage 1827   WRO

TALMAN, William architect
Controller of Works to William III, architect of Chatsworth, of other
palaces, and of alterations to Hampton Court Palace. Born West Lavington,
where he had property. Associate and assistant of Sir Christopher Wren.
Lived at one time at Eastcott, Easterton. Said by ML to have also lived at
Paxcroft Farm, Hilperton.   VCH vol 10 p180-1   VCH vol 7 p199   DNB
1. URCHFONT. Manor. (Ascription. East front - rebuilt in 1690s - likely)
   S
2. TROWBRIDGE. Fore Street, Lloyd's Bank (ascription - same facade as
   centre of Dyrham Court, Glos.)   ML

TARRING & WILKINSON architects
1. SALISBURY. Fisherton Street, Congregational Chapel (now United Reformed
   Church) 1879   P

TATE, N architect
1. DONHEAD ST ANDREW. Restoration work 1875   WRO

TATHAM, Charles Heathcote architect 1772-1842
Associated with Henry Holland, and the Regency style.   C
1. WINTERSLOW. Roche Court 1804   P
2. WHITEPARISH. Cowesfield House, enlarged for Sir Arthur Paget before
   1825   C

TAYLOR, James stone mason and builder of Trowbridge Road, Bradford-on-Avon
Harrods Directory 1865
1. BRADFORD-ON-AVON. Frome Road, Prospect House 1860s   WT 26.1.1996

TAYLOR, John mason of Hilperton Lane, Trowbridge in 1842   ETD
1. TROWBRIDGE. Victoria Road, The Cedars 1858-62   SP
2. TROWBRIDGE. Victoria Road, Rodwell Hall 1859 (architects Manners and
   Gill)   SP

TAYLOR, Sir Robert sculptor & architect 1714-88
Neo-Palladian of the Burlington School. Knighted when Sheriff of London,
1782-3.   F,H & P
1. CHUTE FOREST. Chute Lodge 1768   MG P
2. SALISBURY. Bishop's Palace (now Cathedral School). A Gothick porch
   c1790   P
3. SALISBURY. Guildhall (Elizabethan Council House destroyed by fire
   1780). New Guildhall designed by Taylor and built with some

alterations by William Pilkington 1788-95.   P
4. WESTBURY. Church monument to W. Phipps, Governor of Bombay (d. 1748)
   JO

TEULON, Samuel Saunders architect 1812-1873
Of French descent and a pupil of George Legg and George Potter. BFC
1. TIDCOMBE AND FOSBURY. Fosbury. Christ Church 1854-6   P
2. ALDERBURY. St Mary 1857-8   P
3. WILCOT. Oare, Holy Trinity 1857-8   WRO   P
4. TIDCOMBE AND FOSBURY. Fosbury House?
5. GREAT BEDWYN. Church Street, cottage?

THOMAS, M Hartland architect of Bristol
1. SWINDON. Southbrook, All Saints 1937   P

THOMAS, P.H architect
1. OAKSEY. All Saints Church, restoration c1934   VCH

THOMPSON, James architect of Melrose, Scotland 1800-1883
Apprenticed to J.B.Papworth. Worked for Nash on Cumberland Terrace.
Published 'Retreats' 1827, collection of picturesque rustic cottage
designs. Work in London in 1830s and 40s. Work for Neeld from 1827 at
least (Norton St Philip School). Work for Neeld eventually involved
rebuilding of dozens of cottages, several schools, rectories, isolated
farmhouses, set of almshouses, two churches and a town hall (Chippenham)
apart from Grittleton House itself. All the following are Joseph Neeld
work.   JO
1. LUCKINGTON. Alderton. Estate village 1831-2   JO
2. LUCKINGTON. Alderton, school, now Church Cottage 1832   JO
3. HULLAVINGTON. School 1832   JO
4. GRITTLETON. Fosse Lodge 1835   JO
5. GRITTLETON. East Foscote Farm c.1835   JO
6. CHIPPENHAM WITHOUT. Lanhill Farm 1840   SP   JO
7. GRITTLETON. House 1842. "It is really a monstrosity" - P
8. LUCKINGTON. Alderton. St Giles 1845. The Vicarage is probably
   Thompson also.   P
9. GRITTLETON. Leigh Delamere, St Margaret. Rebuilding 1846   P
10.GRITTLETON. Church, Rectory, almshouses c.1846   JO
11.GRITTLETON. Sevington, school 1847   P
12.CHIPPENHAM. Town Hall 1848   P
13.GRITTLETON. Sevington, cottages by school   JO
14.GRITTLETON. Manor Farm remodelled   JO

THOMPSON, J.J architect
Son of James Thompson?   JO
1. GRITTLETON. Grittleton House. Exhibited design for carriage porch
   at Royal Academy 1863   JO

THORNTON, T architect of Salisbury
Related to Thomas Thornton, plumber, glazier and painter of Downton in 1830
and 1842? ETD
1. WINTERBOURNE. Winterbourne Gunner. Vicarage altered or rebuilt 1854 JO

THURSTON, Simon architect of London
1. WINTERSLOW. Parsonage 1847   WRO

TITCOMBE family carpenters and builders of High Street, Avebury. Albert and George, brothers, early C20. Worked with Shipway (masons) and Paradise (ironworkers) families.   SA
1. AVEBURY. St James's Church. Lych-gate 1899   SA

TOWNSEND, architect
1. CLYFFE PYPARD. Thornhill Manor Farm. New wing built 1723-4 VCH

TOWNSEND, Robert architect of Durrington
Practice at The Studio, Bulford Road, Durrington.
1. WILTON. Royal Carpet Factory. New building behind the main one 1957 P
2. MILTON LILBOURNE. Cobbett's Way. House for Mr and Mrs Anstead 1976 P

TRAVERS, Martin stained glass window designer Died before 1950.
1. SWINDON. Christ Church, window after 1927   JB

TROTTER, A.P architect of Teffont
1. CORSLEY. St Margaret. Restoration work 1937   WRO
2. HEYTESBURY. SS Peter & Paul. Restoration work 1937   WRO
3. CHITTERNE. All Saints. Restoration work 1939   WRO

TROUP, F.W architect
1. DEVIZES. War Memorial   JO

TROW, H.A architect
Practice at 42 High Street, Wootton Bassett (Kelly's 1911, 1915)

TUBBS, Cyril B architect of Reading, Berks.
Of practice WEBB & TUBBS.
1. FROXFIELD. Parsonage 1882   WRO

TURNER, Thomas builder and brickmaker of Swindon
Proprietor of brickworks in Drove Road. JB Living at Grove House 1895. VCH
1. SWINDON. Drove Road. Turner's house demonstrating brickwork 1871 JB
2. SWINDON. Turner Street (architect J.J.Smith) 1894   JB
3. SWINDON. Hunt Street, terrace of 1895 (architect J.J.Smith) JB

TURNER, J architect
1. SALISBURY. Fisherton Anger. Cemetery   JO
2. SALISBURY. Fisherton Anger. Fisherton House (Chapel, library & concert
   room)   JO

TYRWHITT, T architect 1874-
1. NORTON. Foxley Grange, reconstructed early 20C   WWA

UPTON, T architect
Practice at The Bungalow, Haynes Road, Westbury (Kelly's 1927)

UNDERWOOD, C of Bristol
1. TROWBRIDGE. Court Street, Home Mills 1862   JO

UNWIN, Sir Raymond architect
1. SWINDON. Laid out Pinehurst estate for Swindon Corporation
   1922-39 period   JB

VALLIS, R.W.H architect of 6 North Parade, Frome, Somerset
1. BRATTON. Court House, repairs 1947 (Holdoway builder)   WRO

VALLIS & BUTTER architects
1. WINTERSLOW. All Saints. Restoration work 1938   WRO

VAUGHAN, Henry builder of Winchester, Hants.
1. ALDERBURY. Workhouse 1837 (architect Hunt)   RG

VIALLS, George of London
1. WARMINSTER. Sambourne, Christ Church. Arcades 1881   P

VOISEY & WILLS
1. WESTWOOD. Vicarage 1877-8   VCH

WADE, F.B architect
1. SWINDON. Aylesbury Street, St John the Evangelist's Church 1883
   WBR

WADLOW, Henry and John timber merchants and builders of Mere
1. MERE. Workhouse 1840 (Scott and Moffatt architects)   RG

WAILES, E, CLAYTON & BELL stained glass designers
1. CALNE. Church of St Mary (Kelly's 1867)   JO
2. CHIRTON. St John's Church, E window and chancel glass   JO
3. BROMHAM. Chittoe. St Mary's Church. E window   JO
4. STERT. St James Church. E window and one lancet   JO

WAIT, Colonel of Sutton Veny
1. LONGBRIDGE DEVERILL. SS Peter & Paul. Restoration work 1936 WRO

WAKEFIELD, Peter architect of Warminster
1. LONGBRIDGE DEVERILL. Crockerton. House on Low Ridge 1956   P

WALDEGRAVE, Samuel builder or architect of Yeovil, Somerset
1. BARFORD ST MARTIN. National School 1853   WRO

WALKER, Thomas architect
WCC County Architect in the 1920s and 30s.
1. ROUNDWAY. County Lunatic Asylum (later Roundway Hospital), detached
   chapel 1937   RCHM

WALKER, William architect, of Shaftesbury, Dorset
1. DONHEAD ST MARY. Charlton. Chapel 1839   P
2. NORTON BAVANT. All Saints. Rebuilding 1839-41   P

WANINER (or WONINER), William H architect of Stockwell Park
1. MONKTON FARLEIGH. School and house 1872   WRO
2. MONKTON FARLEIGH. Parsonage 1872   WRO

WARD R.J
Designed railway buildings in S. Wiltshire and elsewhere. Talk (D. Hyde?) WANHS 1990.
1. DEVIZES. Station (demolished)
2. MARLBOROUGH. Station (though signed by Brunel)
3. MALMESBURY. Station (though signed by Brunel)

WARD, Thomas painter
1. WILTON. Wilton House. Ceiling painting 1816. (Possibly the painter active between 1819 and 1840.) RCHM

WARD & HUGHES stained glass designers
1. MELKSHAM. St Michael. S Chapel SW window 1884 JO

WARD & NIXON stained glass designers
1. CASTLE COMBE. St Andrew's Church. E, W, Nave E and N chapel E windows P JO

WARING & BLAKE of London
1. HAM. Rectory. New W. wing 1864 VCH

WARRE, Edmund architect
1. WILTON. Wilton House. Rebuilding of N facade. Conversion of Wyatt's library into a drawing room 1913 RCHM P

WARREN, E.P architect
1. EBBESBOURNE WAKE. St John. Restoration work 1899 WRO
2. WEST LAVINGTON. Littleton Pannell, A Becketts 1904 P

WARRINGTON, stained glass designer
1. TROWBRIDGE. St James Church. Window given by Manners & Gill, architects. JO

WATERS, John surveyor of Salisbury
1. STEEPLE LANGFORD. School, site plan 1858 WRO

WATTS, Richard Pocock timber merchant and builder of London Road, Chippenham
1. CHIPPENHAM. Workhouse, built to design by Creeke 1859 RG

WEAVER, Henry architect
Agent for Poynder estate in late 1840s and designed Beversbrook Farmhouse, Hilmarton, for himself as agent and as agent's house. JO
Practising from Beversbrook House, 'Calne' in 1857. Later practice at 31 Long Street, Devizes (Kellys 1875, 1880). Sent business letter from Devizes 2.6.1868. WRO County Surveyor by 1865 RCHM.
1. CALNE WITHOUT. Bowood Estate. Pillars Lodge, Mile End c1848 JO
2. CALNE WITHOUT. Calstone. Parsonage 1849 WRO
3. MALMESBURY WITHOUT. Corston. Vicarage work in 1849 JO WRO
4. HILMARTON. St Lawrence. Restoration c1850, including rebuilding of S porch. P VCH
5. HILMARTON. Hilmarton Lodge (was shooting lodge) c1850-60
6. HILMARTON. School 1851 VCH
7. STANTON ST BERNARD. Little Farmhouse, alterations and extensions

       1851 (Vicars of Stanton had lived there since 1823.)   VCH
8. CHIPPENHAM. St Paul's School 1857   WRO
9. ROUNDWAY. County Lunatic Asylum. New wing 1865   RCHM
10. ALLCANNINGS. All Saints Church, chancel 1867   P
11. SEEND. Holy Cross. Restoration work (not present building). WRO
12. ETCHILHAMPTON. St Andrew. Restoration 1868-9 (chancel rebuilt, vestry
    added).   VCH
13. ROWDE. St Peters. School 1868   WRO
14. WEST ASHTON. Work for Rood Ashton estate. Alterations to estate
    cottages (probably at Heath Hill) and to lodge, and discussion of
    addition of small mortuary chapel to S side of chancel of West
    Ashton Church   1868   Letter, WRO
15. WILSFORD. St Michael. Porch 1869   P
16. BROAD TOWN. Parsonage 1870   WRO
17. HILPERTON. Additions to school 1870   WRO
18. DEVIZES. Long Street. Museum - centre part 1872   P
19. WINTERBOURNE. Winterbourne Gunner. Parsonage 1872   WRO
20. CHERHILL. Parsonage 1872   WRO
21. BERWICK BASSETT. Parsonage 1872   JO   WRO
22. RUSHALL. Parsonage 1873   WRO
23. MARDEN. Parsonage 1874   WRO
24. DEVIZES. Southbroom. St James. Infant School 1874   WRO
25. EASTERTON. School 1875   WRO
26. POULSHOT. Parsonage 1875   WRO
27. UPTON LOVELL. Parsonage 1875   WRO
28. PATNEY. Two cottages on glebe land S of the church 1875   VCH
29. PATNEY. Parsonage 1876   WRO
30. EASTERTON. Parsonage 1876   WRO
31. ENFORD. Parsonage 1876   WRO
32. PATNEY. St Swithin. Restoration 1876-8   VCH
33. SEEND. Parsonage 1877   WRO
34. HILMARTON. Poynder almshouses 1878   VCH
35. HILMARTON. Estate cottages in the village e.g at Snow Hill. (Weaver was
    agent for the Hilmarton estate for a time)   VCH
36. COMPTON BASSETT. St Swithin. Restoration work 1880   WRO
37. COLLINGBOURNE KINGSTON. St Mary. Restoration work 1880   WRO
38. BROUGHTON GIFFORD. Vicarage c1880   JO   WRO
39. LITTLE BEDWYN. St Michael. Restoration work 1882   WRO
40. BROUGHTON GIFFORD. St Mary. Restoration work 1883   WRO
41. SOUTH WRAXALL. St James. Restoration work 1884   WRO
42. CALNE WITHOUT. Studley. Old Road. Laburnum Cottage/Willow Cottage.
    JO
43. COLLINGBOURNE KINGSTON. Vicarage. Alterations   JO
44. LITTLE BEDWYN. Vicarage. (Undated) works   JO   WRO
45. CALNE WITHOUT. Black Dog Halt station house? WANHS talk (D. Hyde?) 1990
46. CORSHAM. Hartham. Home Farm?   JO

WEAVER & ADYE architects of Bradford-on-Avon and Devizes
H.Weaver of Devizes took C.S.Adye of Bradford as partner at the end of his
career. Adye (q.v.) later became County Surveyor for Wiltshire.   JO
1. TROWBRIDGE. Union Street, St James Hall 1880 (1882 WT) for Mr R. Rodway.
   JO
2. SOUTH WRAXALL. St James. Restoration work 1881   WRO
3. SOUTH WRAXALL. Vicarage c1882   JO

4. BRADFORD-ON-AVON. Silver Street, brewery buildings, castellated
   store of 1884   JO
5. ROWDE. St Matthew. Restoration work 1886   WRO

WEBB, Sir Aston architect   1849-1930
Particularly successful at public buildings, most in a Baroque style e.g. V
& A Museum 1891- , Royal Naval College, Dartmouth, 1899-1904, Birmingham
University, 1906-9, facade of Buckingham Palace, 1913.   F,H & P LB
1. STOURTON WITH GASPER. Stourhead House. Reconstruction after 1902 fire
   (with Doran Webb)   P
2. MARLBOROUGH. College, Field House 1910-11   P
3. MARLBOROUGH. College, Footbridge between gymnasium and North Block
   1910-11   VCH

WEBB, Christopher architect
1. ODSTOCK. Nunton, St Andrew. Restoration work 1936   WRO
2. CHISELDON. Holy Cross. Restoration work 1939   WRO

WEBB, E Doran architect
Practice at 58 High Street, Salisbury (Kelly's 1889, 1895); at 75 New
Street (Kelly's 1899); at The Close Gate (Kelly's 1907); at 37 New Street
(Kelly's 1911, 1915)
1. SALISBURY. St Osmund. North aisle 1894   RCHM
2. SALISBURY. St Martin. Restoration work 1896   WRO
3. STOURTON WITH GASPER. Stourhead House. Reconstruction after 1902
   fire (with Sir A.Webb)   P
4. SWINDON. Groundwell Road, Holy Rood 1905   P
5. WEST TISBURY. Newtown, Church 1911   P

WEBB, John architect 1611-1672
Pupil and nephew by marriage of Inigo Jones, but without his imagination
and originality. King Charles Building at Greenwich Hospital (1662-9) is
the best of his buildings to survive.   F, H & P
1. WILTON. Wilton House. Advised de Caus with Jones during 1630s alteration
   and enlargement. Rebuilt S range 1649-52 after 1647 fire. Drawings
   for interior by Jones & Webb survive.   RCHM   P
2. AMESBURY. Amesbury Abbey, before 1660. Described as a "triumph in
   country house design". (Radical rebuilding by Hopper in 1834.) P RCHM

WEBB, Philip architect & furniture designer 1831-1915
His first job was Red House, Bexleyheath, Kent (1859) for William Morris.
Almost exclusively a domestic architect. He and Norman Shaw are the best
of the English Domestic Revival. Ceased practising in 1901.   F, H & P
Few of his houses remain standing, let alone unaltered. LB
1. EAST KNOYLE. Clouds. Designed 1879, built 1881-6; burnt out 1889,
   rebuilt 1889-91   P
2. EAST KNOYLE. St Mary. Restoration of tower 1893   P
3. WILSFORD CUM LAKE. Lake House. An Elizabethan house restored by Detmar
   Blow in 1898 "with the council of Philip Webb".   P

WEBB & SUTTON architects of Reading, Berks
1. RAMSBURY. Holy Cross. Restoration work 1909   WRO

WEEDON, H.W and PARTNERS architects of Birmingham
Harry Weedon. Later the WEEDON PARTNERSHIP.
1. SWINDON. Pressed Steel Co. factory 1957-62   P
2. SWINDON. Civic Offices - extensions (with J. Winter) 1975 P

WESTMACOTT, Richard sculptor
Monuments in various Wiltshire churches. P
1. WILTON. Wilton House. Oversaw completion of cloisters 1815. Door replacements, new bridge over Wylye and loggia in Italian garden 1826. Re-set marble fountain.   RCHM

WHALL, Christopher stained glass designer
1. CHIPPENHAM. St Andrew's Church, S chapel window   JO

WHEELER, William builder
1. LYDIARD MILLICENT. Additions to school and house 1858   WRO
2. LYDIARD MILLICENT. School, new classroom 1866   WRO
3. LYDIARD MILLICENT. School, further additions 1870   WRO

WHEELER, William architect
1. ALDBOURNE. St Michael. Restoration work 1935   WRO

WHITE, Benoni Thomas architect 1808-51
Known as Thomas. Son of J.Benoni White (qv). JO In partnership as Young and White (qv). WRO
1. AVEBURY. Parsonage 1841   WRO
2. CALNE WITHOUT. Calstone. Parsonage 1842   WRO

WHITE, J Benoni architect of Devizes 1784-1833
This may be Benoni White junior who was a builder at Leg of Mutton Street, Devizes in 1822. ETD Correspondence with J. Peniston (qv) 1825. PL
1. BROMHAM. Rowdeford House, alterations for Wadham Locke supervised by J. Peniston 1825   PL
2. WEST LAVINGTON. Parsonage 1830   WRO

WHITE, James builder
1. CALNE WITHOUT. Bowood House. Work 1765-c1770   WAM 41

WHITE, William architect of London
1. WESTBURY. Westbury Leigh, Holy Saviour, designed 1851 (though not built till 1876-80)   P
2. RAMSBURY. Axford. St Michael 1856   P
3. CHUTE. Schools 1857-8   P
4. CHUTE. Vicarage 1860   JO WRO
5. MARLBOROUGH. College, Master's Lodge, porch c1862-3; sick house 1862-3; Barton Hill and Elmhurst 1862-3   P
6. PRESHUTE. Rectory. Extension c1863   VCH
7. FITTLETON. All Saints. Restoration work 1878   WRO

WILCOX, John master carpenter
1. DAUNTSEY. Rectory 1829-33   VCH

WILKES (WICKS), John glazier
1. SEMINGTON. Whaddon House and stable. Extensive work 1673, 1677 and 1679 LP

WILKINS, W.H architect
1. TROWBRIDGE. Studley, St John 1858   P

WILKINSON, George architect of Witney, Oxon 1814-1890
Designed workhouses in Oxfordshire and at Dorchester. In 1838 or 1839 he went to Ireland, served as architect of Irish Poor Law and designed many workhouses there. See also DARLEY & WILKINSON.   RG
1. DEVIZES. Workhouse 1836 (built by Young and White)   RG
2. PURTON. Cricklade and Wootton Bassett Union Workhouse 1837 (mostly demolished) RG

WILKINSON, W architect of Oxford
1. HORNINGSHAM. Longleat. Stalls Farm (a model farm) 1859   P
2. HORNINGSHAM. Longleat, Park Farm (a model farm) 1860   P
3. EBBESBORNE WAKE. Fifield Bavant, Farm; entry in Arch.Eng.
   & Building Trades Directory, 1868   JO
4. HIGHWORTH. Sevenhampton. Two lodges for Warneford Place 1872   P

WILKINSON AND MOORE architects of Oxford
1. DILTON MARSH. Dilton Court, farm buildings, 1842 or later.
   Owner's plans

WILLCOX, W.J architect of Bath
(See also Wilson and Willcox.)
1. CHIPPENHAM. Derriads Lane, Derriads, alts. c1905   JO

WILLES, Thomas architect, carpenter and builder of Marlborough
At the Green in 1842 ETD
1. MARLBOROUGH. Marlborough Workhouse (as builder) 1837   RG
2. MARLBOROUGH. St Mary, Parsonage 1839   WRO

WILLIAMS, Henry architect of Corn Street, Bristol
1. BRADFORD-ON-AVON. 7 & 8 Silver Street, reconstructed 1876 HF

WILLIAMS-ELLIS, Sir Clough (see **ELLIS**)

WILLIS, E architect
1. STAVERTON. St Paul. Restoration work 1936   WRO

WILLIS, Frank of Bristol
1. WESTBURY. Warminster Road, Leighton House, late C19 additions. DOE

WILLOUGHBY, John Bean architect of London
1. BERWICK ST JOHN. Parsonage 1880   WRO

WILMSHURST, Thomas stained glass designer
In partnership with **F.W.Olliphant** from 1853-5.
1. GRITTLETON. Leigh Delamere, St Margaret's Church, "in the most strident colours" - P   JO

WILSON, James architect of Bath 1816-1900
Designed a lot of villas on the Weston and Lansdown sides of Bath. JO
1. BISHOPSTROW. Vicarage, plans for alts., undated   JO
2. COLERNE. Lucknam Park. Chippenham Lodge looks like a c1860 design of his.   JO
3. WESTWOOD. Iford Manor. Plans for alterations to service wing and rear staircase, undated c1850-60? RIBA Drawings collection.   JO

WILSON & FULLER architects of London and Bath
1. WESTWOOD. Plan of churchyard etc (re school)   WRO

WILSON & WILLCOX architects of Bath and London
1. SWINDON. Town Hall. Addition of Italianate tower, etc. to market house 1866   VCH
2. MELKSHAM. High Street, Methodist Chapel 1872   JO
3. COLERNE. Designs for Walmesley of Lucknam Park include Euridge Farm Cottages, 1876; restoration of St John the Baptist church 1875-6; Euridge Manor Farm 1877; possibly did water tower at Lucknam Park. Additions to village school 1893 and memorial column to Walmesley in village may be by W.J.Willcox. Surviving drawings are in the Crozier-Cole collection, Brit. Architectural Lib.   JO
4. COLERNE. The Manor House dated 1689, but much rebuilt 1900 JO
5. DEVIZES. Maryport Street. Oddfellows Hall. 1st prize design (built?) JO

WILLCOX W.J architect of Bath
See also WILSON & WILLCOX.
1. CHIPPENHAM. Derriads Lane, Derriads, altered c1903   JO

WILTSHIRE, George builder of Swindon
Advertising 1882 in Deacon's Gazetteer.

WITHERS, R.J architect
1. BUTTERMERE. St James. Rebuilt 1855-6   P
2. SUTTON BENGER. Draycot Cerne. Draycot House, NW wing 1864 VCH
3. AVEBURY. St James. Chancel rebuilt 1879   P
4. DILTON MARSH. Vicarage   JO

WOOD, John the elder town planner and architect 1704-1754
Palladian. Builder of C18 Bath.
1. BRADFORD-ON-AVON. Belcombe Court. Wing added 1734   P
2. MONKTON FARLEIGH. Manor House. Additions and rebuilding   VCH
3. LACOCK. House at Bowden Hill c.1744 (later rebuilt at 24 High Street, Chippenham, dismantled c1932 and re-erected at Zion Hill, Bath by 1946. Now used as part of Kingswood School.)   TM

WOOD, John the younger architect of Bath 1728-1787
Acted as his father's assistant as a young man, but took over as the leading architect in Bath on his father's death in 1754. The Royal Crescent is his best work.   C
1. DOWNTON. Standlynch. Trafalgar House. Wings added to Roger Morris's 1733 house, 1766   P
2. SALISBURY. Fisherton Street, General Infirmary, centre 1767-71

   P
3. CHIPPENHAM. Hardenhuish, St Nicholas 1779   P

WOOD, Joseph architect of Bristol
1. WOOTTON BASSETT. Parsonage 1866   WRO

WOODMAN, James carpenter
1. STANTON ST QUINTIN. Upper Stanton Farm, repairs 1836   EG

WOODMAN, John mason of Chippenham
Bought windmill and two cottages at Kington St Michael in 1818. MW

WOODMAN, John plasterer, slater and bricklayer of The Green, Devizes 1830 ETD

WOODMAN, Thomas mason of Turleigh, Winsley
1. BRADFORD-ON-AVON. 10 Belcombe Place. Bought plot in 1825 with H.Rushton (qv) and house there let by 1828   PMS

WOODMAN, Richard slater
1. DAUNTSEY. Rectory 1829-33   VCH

WOODMAN, W.H architect
1. GREAT CHEVERELL. St Peter, restoration 1868   P

WOODYER, Henry architect of Grafham House, Guildford, Surrey
1. SALISBURY. Bedwin Street, St Edmunds Church School 1860   P
2. BERWICK ST JOHN. St John. Restoration 1861   P
3. COMPTON BASSETT. St Swithin. Chancel, chancel chapels and N porch, 1866   P
4. SALISBURY. Fisherton Anger Church School 1867   P
5. DEVIZES. Market Place fountain 1879 (attribution)   P
6. LACOCK. Cantax Hill, The Old Rectory. Possibly by Woodyer JO

WREN, Sir Christopher architect 1632-1723
Born at East Knoyle, Wilts. First work was chapel, Pembroke College Cambridge 1663. Best known for his London churches after 1666 fire and many public buildings. CH
1. SALISBURY. Cathedral. 1668 - report and recommendation for long-term restoration measures. Tie-rods put in tower, first by Wren (later by Price and Scott)   P
2. SALISBURY. Matron's College. Possibly designed or approved by Wren, 1682   P
3. PITTON and FARLEY. Farley. All Saints 1689-90 (probably built by Alexander Fort, but Wren may very well have helped design the church).   P
4. WARMINSTER. Church Street, Lord Weymouth's Grammar School. Doorcase, which came from Longleat, was designed by Wren.   VCH

WYATT, Benjamin Dean architect 1775-c1855
Eldest son of James Wyatt. Took up architecture in 1809. GW Surveyor of Westminster Abbey 1813-1827. Very competent architect, but not much imagination. C
1. DEVIZES. Market Cross 1814 (with L.J.Abington)   P

WYATT, James architect 1746-1813
Classical to Greek Revival in his large houses, then neo-Gothic (Fonthill
Abbey). Ruthless restorer and "improver" (Salisbury, Durham, Hereford) -
earned him the name of Wyatt the Destroyer.   F H & P
1. DEVIZES. New Park (later called Roundway House) 1780 (demolished 1955)
   B
2. SALISBURY. Cathedral. "Restoration" 1789   P
3. CORSHAM. Hartham Park 1790-5   P
4. SALISBURY. Bourne Hill, Council House (formerly College House). A C15
   porch from the Cathedral re-erected after removal by Wyatt in 1791. P
5. ODSTOCK. Nunton. Bodenham. New Hall. Original house, now replaced,
   attributed to Wyatt 1792   P
6. FONTHILL GIFFORD. Fonthill Abbey. Beckford's Gothicizing of the old
   building 1793-1807   P
7. WILTON. Bulbridge House. Additions 1794   P
8. WROUGHTON. Salthrop House. Main block c1795 (style of J. Wyatt)
   VCH
9. LACOCK. Bowden Park 1796   P
10. WILTON. Wilton House. Widespread alterations 1801-12. "An unfortunate
    attempt to transform the most important 17thC classical house into a
    medieval abbey." Cloisters were most effective contribution to plan but
    poor supervision of his work in general led to problems. RCHM P
11. ODSTOCK. Longford Castle. Designs for remodelling (David Asher
    Alexander did the work 1802-17)   P

WYATT, Sir Matthew Digby architect 1820-1877
Born at Rowdeford House, Bromham and went to school at Devizes. JMR
Still at Rowdeford in 1850 P. He was a leading figure in the architectural
profession, Secretary to the 1851 Exhibition, RIBA Gold Medallist 1866 and
designer of the India Office parts of the Foreign Office complex. JO
Brother of T.H.Wyatt. Paddington Station Hotel in French Renaissance style
is his best-known work. LB
1. ROWDE. St Matthew. Font 1850   P
2. ALLCANNINGS. All Saints Church. Elaborate chancel 1867. P says by
   Weaver but JO says M.D.Wyatt may well have had a hand in it (Kelly
   1867).

WYATT, Thomas of Rowdeford House, Bromham
Living at Rowdeford in 1793-8. ETD One of the commissioners for the Bishops
Cannings Enclosure Award in 1812 so possibly a surveyor.
1. BROMHAM. Rowdeford House. Rebuilt 1812. T. Wyatt may have designed
   the new house   BB

WYATT, Thomas Henry architect 1807-1880
His father Matthew lived at Rowdeford House, Bromham. JMR Wilton Church
(see Wyatt and Brandon) is his chef d'oeuvre. Designer of many Gothic
churches, for a time in partnership with David Brandon. Obituary in The
Builder 14.8.1880 JO
1. DEVIZES. Northgate Street, Assize Court 1835. "A pure Greek portico"-
   JO
2. MELKSHAM WITHOUT. Shaw. Christ Church 1836-8 (mostly rebuilt by Ponting
   1905)   P
3. CHOLDERTON. St Nicholas. New church 1841-50 with Thomas Mozley and
   mostly at expense of Mozley.   VCH

4. CODFORD. St Mary. Restoration work 1843 but his arcade later replaced. P
5. DILTON MARSH. Holy Trinity 1844   P
6. MONKTON FARLEIGH. St Peter. Rebuilding 1844
7. BROMHAM. Chittoe. St Mary 1845   P
8. MELKSHAM. St Michael, restoration 1845
9. WOODFORD. All Saints 1845 (not the tower)   P
10. COOMBE BISSETT. St Michael, restoration 1845   JO
11. MERE. St Michaels Church, restored 1845   JO
12. WROUGHTON. Sts John and Helen, restored 1846   JO
13. WESTBURY. All Saints. Some restoration 1847   P
14. BROUGHTON GIFFORD. Rectory 1848   P
15. ROUNDWAY. Wiltshire County Pauper Lunatic Asylum (later Roundway Hospital) 1849-51.   Italianate   RCHM
16. WOODBOROUGH. St Mary Magdalene. Rebuilding 1850 (chancel), 1861 (nave and aisle)   P
17. UPTON SCUDAMORE. Vicarage altered or rebuilt 1850   JO
18. WILTON. Wilton House. Minor alterations before 1851   RCHM
19. SAVERNAKE. Cadley, Christ Church 1851. "Dull...(except) naughty W front" - P
20. DOWNTON. Charlton, All Saints 1851   P
21. WILSFORD CUM LAKE. St Michael, mostly Wyatt 1851   P
22. HILPERTON. St Michael 1852 (tower is older)   P
23. SALISBURY. Fisherton Anger, St Paul 1851-3   P
24. SALISBURY. Harnham, school c1852   JO
25. ALDERBURY. Vicarage 1852   RA
26. SALISBURY. Harnham, All Saints 1852-4   JO P
27. BURBAGE. Parsonage 1853 (Builder: 14.8.80) JO (T.H.Wyatt obit.)
28. GREAT BEDWYN. St Mary, restoration 1853-4
29. BURBAGE. All Saints, restoration 1854, S aisle (Stanton Memorial aisle) added 1876   JO
30. PRESHUTE. St George, exterior (except tower) 1854; inside arches 1854   P
31. ODSTOCK. Nunton, St Andrew 1854-5   P
32. BURBAGE. School 1854   WRO
33. SHREWTON. St Mary. Restored and enlarged 1854. VCH Tower, arcades and chancel arch are not his. P
34. WARMINSTER. Corn Market 1855 (used as a garage and shops in 1962) VCH
35. DEVIZES. Bath Road, County militia stores, 1856 (part converted in 1879 into the headquarters of the county police, and continued so until the removal to new premises in London Road, 1962)   VCH
36. GRITTLETON. Littleton Drew, All Saints, rebuilt 1856   JO
37. BAYDON. Glebe House 1857-8   VCH
38. MARKET LAVINGTON. St Mary 1857   WRO   JO
39. LAVERSTOCK. St Andrew 1858. Builder 24.7.1868   JO
40. BISHOPSTONE S. St John, restored 1858   WRO   JO
41. BURCOMBE WITHOUT. St John, N aisle 1858-9
42. BISHOPS CANNINGS. St Mary, Restored 1860   JO
43. BOYTON. St Mary. Restoration 1860   P
44. COMBE BISSETT. Homington, St Mary, restoration 1860   JO
45. CHOLDERTON. Upper Farm (later Drybrook Cottage) c1860   VCH
46. SALISBURY. Bemerton, St John 1860-1 (for Pembrokes of Wilton) P
47. SAVERNAKE. Savernake Forest, St Katherine 1861 (for Marchioness of

    Ailesbury)   P
48. NORTH BRADLEY. St Nicholas. Mostly rebuilt 1862   P
49. SOUTH NEWTON. St Andrew, rebuilding 1861-2   VCH   P
50. SUTTON MANDEVILLE. All Saints, mostly Wyatt, 1862   P
51. MARLBOROUGH. SS Peter & Paul. Chancel remodelled 1862-3   P
52. CHITTERNE. All Saints 1863   P
53. FOVANT. St George's Church, restoration 1863   JO
54. GREAT WISHFORD. St Giles, rebuilding 1863-4   P
55. CODFORD. St Peter, chancel 1864   P
56. STEEPLE LANGFORD. Little Langford, St Nicholas, rebuilt 1864 P
57. ALVEDISTON. St Mary, mostly Wyatt, 1866   P
58. WINTERSLOW. All Saints, exterior, 1866   P
59. BOWER CHALKE. Holy Trinity, South aisle & chancel 1866   P
60. FONTHILL GIFFORD. Holy Trinity 1866. For Marquess of Westminster. P
61. SEDGEHILL AND SEMLEY. Semley. St Leonard, rebuilt 1866-76 DOE
62. WINTERBOURNE. Winterbourne Earls, St Michael 1867-8   P
63. HINDON. St John the Baptist, rebuilding 1870-1   VCH
64. WARMINSTER. Sambourne, Christ Church, chancel 1871   P
65. SOPWORTH. St Mary Church, restored 1871   JO
66. SALISBURY. Bemerton. School 1871   WRO
67. OGBOURNE ST GEORGE. Church of St George, restored 1873   JO
68. UPAVON. St Mary, chancel restoration 1875   P
69. FONTHILL BISHOP. All Saints. Restoration 1879
70. WEST ASHTON. Rood Ashton House. West Ashton Lodge at S entrance.
    DOE
71. BRATTON. Church of St James, chancel in association with G.G.Scott
    (Kelly 1867)   JO
72. CLARENDON PARK. Clarendon Park, ballroom and service wing added
    for Sir F.Bathurst   JO
73. GRITTLETON. Littleton Drew, Rectory. Possibly by Wyatt   JO
74. SOUTH NEWTON. St Andrew's Church, vicarage   JO

WYATT, Thomas Henry and BRANDON, David architects
(See separate entry for Wyatt.)
1. CALNE WITHOUT. Derry Hill, Christ Church 1839-40   P
2. CHOLDERTON. St Nicholas 1840-50. Medieval roof brought from a
   warehouse on the quay at Ipswich.   JO   P
3. WILTON. St Mary & St Nicholas, 1841-5. (Builders Jones Bros.)
   "One of the few outstanding examples of Italian Romanesque revival
   architecture in England" - LB
4. LONGBRIDGE DEVERILL. Crockerton, Holy Trinity Church 1843 P
5. WORTON. Christ Church 1843   P
6. WYLYE. St Mary, restoration work 1843   WRO
7. SHREWTON. Maddington. St Mary's Church, restoration 1843-53 VCH
8. HORNINGSHAM. St John (not the tower) 1844   P
9. NEWTON TONEY. St Andrew 1844   P
10. WYLYE. St Mary's Church 1844-6   VCH
11. TILSHEAD. St Thomas, restoration work and alterations 1845 VCH WRO
12. EAST KNOYLE. St Mary, rebuilt 1845   VCH
13. WEST ASHTON. St John the Evangelist 1846   P
14. WARMINSTER. National School 1846   WRO
15. WESTBURY. All Saints Church, restored 1847   JO
16. FONTHILL GIFFORD. Fonthill Abbey. A pavilion forming the W service wing
    of the old house was incorporated into a house designed for James

Morrison, 1848-9   P
17. WINTERSLOW. All Saints, restoration work 1849   WRO
18. SALISBURY. The Close, Diocesan Training School (King's House and Sub-
    chantry) 1849   WRO
19. CHILMARK. Church of St Margaret   JO
20. FONTHILL GIFFORD. Fonthill House (since demolished) 1848-9 JO
21. WOODBOROUGH. Vicarage built 1850   WRO   JO

WYATVILLE, Sir Jeffry architect 1766-1840
Nephew of James Wyatt, under whom he trained. Competent classical architect, but specialised in neo-Gothic and "Tudor Collegiate" mansions. Best known for his remodelling of Windsor Castle for George IV, 1824-37. Changed his name to Wyatville when he began work on the Castle. Knighted 1828.   F, H & P
1. STOCKTON. Stockton House, staircase put in c1800 by Wyatville (?) P
2. HORNINGSHAM. Longleat House, interior remodelling 1801-11 (ceiling above grand staircase 1806-13) P
3. HORNINGSHAM. Longleat, County Cottage 1803, cottage orne MB
4. WEST ASHTON. Rood Ashton House for R.G.Long, 1808 (larger part demolished in 1970s)   DOE   P
5. DINTON. Philipps House 1812-16 (75 drawings at estate office GW) P
6. WILTON. Wilton House. Designs for ground floor breakfast-room in S range 1814   RCHM

WYVERN DESIGN GROUP architects of Swindon and Devizes
Formed in 1952 from Edwards & Webster (qv) and R.J.Beswick & Son (qv). PE ADK Many records at WRO.

YOUNG, John and WHITE, Benoni Thomas surveyors and builders of Devizes
Young was a builder at High Street, Devizes in 1822. ETD In partnership with White (q.v.); Young & White in 1842-3 of Bridewell Street and the Green, Devizes ETD. Kelly's 1848
1. DEVIZES. Workhouse 1836 (architect G. Wilkinson)   RG
2. AVEBURY. National School 1842   WRO
3. COMPTON BASSETT. Parsonage 1842   WRO
4. BREMHILL. National School 1846   WRO

Compton Bassett Parsonage (WRO)

-110-

Mr. C. S. Adye, M.S.C.S.

Mr. Fred Bath.

Mr. G. L. W. Blount.

Mr. A. C. Bothams.

Mr. Harold Brakspear.

Capt. J. W. Brooke.

Mr. Edward Drew.   Mr. E. C. Isborn.

Mr. C. W. Gater, J.P.   Mr. W. A. H. Masters.

Mr. H. E. Nicholls.    Mr. W. R. Osborne.

Major A. J. Randell, J.P., V.D.    Mr. Edward Stockwell.

PARISH INDEX

This index covers the location of buildings in the first index and does not include the residence or business location of the architects and building craftsmen. Where several names have the same initial they are not distinguished as it should be possible to locate the correct reference(s) quite quickly.

ALDBOURNE
Brakspear, H., Butterfield, W., Dyer, C., Gimson, H.M., Mason, G., Pinchard, B., Ponting, C.E., Pugin, A.W.N., Rendell, F. and Sons, Wheeler, W.

ALDERBURY
Gambier Parry, S., Hale & Son, Hall, H., Hunt, E., Nichols, G.B., Teulon, S.S., Vaughan, H., Wyatt, T.H.

ALLCANNINGS
Brakspear, H., Weaver, H., Wyatt, M.D.

ALLINGTON
Fisher, F.R., Harding, J., Peniston, J., Shurmer, T.M.

ALTON
Aylmer, G., Gimson, H.M., Ponting, C.E.

ALVEDISTON
Miles, T.B., Wyatt, T.H.

AMESBURY
Blow, D.J., Bothams, Brown & Dixon, Butterfield, W., Chambers, W., Cole, J.J., Curtis, H., Fildes, G., Fleming, Flitcroft, H., Flooks, J.H., Hopper, T., Noyes & Green, Osgood, I., Ponting, C.E., Privett, W., Rendell, F. and Sons, Scott, G.G. & Moffatt, W.B., Webb, J.

ANSTY
Arnold, W., Harding, J. & Son.

ASHTON KEYNES
Butterfield, W., Lansdown & Shopland.

ATWORTH
Brakspear, H., Goodridge, H. E.

AVEBURY
Brakspear, H., Button, Ponting, C.E., Rendell, F. and Sons, Shipway family, Titcombe family, White, B.T., Withers, R.J., Young, J. and White, B.T.

BARFORD ST MARTIN
Fisher, F.R., Stent, T., Waldegrave, S.

BAYDON
Messenger, H., Reeve, J.A., Street, G.E., Wyatt, T.H.

BEECHINGSTOKE
Gabriel, S.B.

BERWICK BASSETT
Rendell, F. and Sons, Weaver, H.

BERWICK ST JAMES
Harding, M.

BERWICK ST JOHN
Bean, W.J., Willoughby, J.B., Woodyer, H.

BERWICK ST LEONARD

BIDDESTONE
Brakspear, H., Rudman, W.

BISHOPS CANNINGS
Brakspear, H., Gimson, H.M., Manning, J., Ponting, C.E., Randell, J.A., Rendell, F. and Sons, Wyatt, T.H.
BISHOPSTONE N
Christian, E., Kempe, C.E., Pace, R.
BISHOPSTONE S
Frith, C.S., Harding, J., Hutchins, T., Lowder, J., Pugin, A.W.N., Wyatt, T.H.
BISHOPSTROW
Champion, W.S., Kempe, C.E., Pinch, J., Wilson, J.
BLUNSDON ST ANDREW
Butterfield, W., Christian, E., Kempe, C.E., Lavers & Barraud, Mantell, E.W.
BOWER CHALKE
Coombs, W., Wyatt, T.H.
BOX
Brakspear, H., Brunel, I.K., Darley, J., Godwin, E.W., Hicks, J.
BOYTON
Butcher, R. & Son, Hardick, W. & Son, Hill, H.L.G., Wyatt, T.H.
BRADFORD-ON-AVON
Adye, C.S., Adye, H.A., Blacking, W.H.R., Brakspear, H., Brunel, I.K., Clutton, H., Deverell, E., Deverell, E. and J., Fuller, T., Gane, R., Gill, J.E., Hardman, J., Hulbert, W.F.E., Irvine, J.T., Jones, D. & Jones, C., Jones, I., Jones, J., Jones, J., Knapp, J., Llewellins & James, Long, J., Long, W., Manners, G.P., Manners & Gill, Marchant, R., Merrick, H., Mitchell, E., Morgan, H.T., Newman, R., O'Connor, M. and A., Ponting, C.E., Pope, R.S., Rolfe & Peter, Rushton, H., Scott, J.O., Silcock, T.B., Spender, J., Stocking, T., Stocking, W., Taylor, J., Weaver & Adye, Williams, H., Wood, J., Woodman, T.
BRATTON
Holdoway, T. & Sons, Hurle, Messenger, H., Pinchard, B., Scott, G.G., Vallis, R.W.H., Wyatt, T.H.
BRAYDON
BREMHILL
Butterfield, W., Hardman, J., Kempe, C.E., Young, J. and White, B.T.
BRINKWORTH
Darley, J. & Sons, Darley, R., Ponting, C.E.
BRITFORD
Street, G.E.
BRIXTON DEVERILL
Harding and Elgar, Messenger, H., Stent, W.J.
BROAD CHALKE
Coombs, W., Harding and Elgar, Stent, W.J.
BROAD HINTON
Baverstock, W.E., Beswick, R.E.E., Ponting, C.E.
BROAD TOWN
Blacking, W.H.R., Campbell, W.H., Kingeston, F., Sage, S., Weaver, H.
BROKENBOROUGH
BROMHAM
Burn, W., Davis, C.E., Gimson, H.M., Manners, G.P., Morris, W., Peniston, J., Ponting, C.E., Rendell, F. and Sons, Slater, W., Slater, W. & Carpenter, R.H., Stevens, E., Wailes, Clayton & Bell, White, J.B., Wyatt, T., Wyatt, T.H.
BROUGHTON GIFFORD

Foley, J.H., Medlicott, W.B., Schmidt, Weaver, H., Wyatt, T.H.
BULFORD
Blount, G.L.W., Ponting, C.E., Rendell, F. and Sons, Sturgess, E.
BULKINGTON
Cundy, T.
BURBAGE
Oliver, E.K., Wyatt, T.H.
BURCOMBE WITHOUT
Harding and Elgar, Wyatt, T.H.
BUTTERMERE
Elliott, S., Withers, R.J.
CALNE
Adam, R, Allom, T., Ashbee, C.R., Bath, F., Blacking, W.H.R., Bowden, F.I., Brakspear, H., Burton, D., Caroe, W.D., Davis, C.E., Green, Lloyd & Son, Morrish, W.J.M., O'Connor, M. and A., Oliver, C.B., Pearson, J.L., Pilkington, W., Robertson, W., Robinson & Kay, Rudman, W., Rutherford, T., Salway, G.S., Slater, W., Smith, C.H., Stent, W.J., Wailes, Clayton & Bell.
CALNE WITHOUT
Adam, R., Barry, C., Button, J., Carlini, Carter, B. and T., Cipriani, G.B., Cockerell, C.R., Comley, G.J.D., Dance, G., Hart, M., Holland, H., Keene, H., Kempe & Tower, Lane, J., Light & Smith, Linnell, J., Methuen, A.P., Oakford, C., Pearsall, T., Ponting, C.E., Powell, I., Rigaud, J.F., Rose, J., Rudman, W., Slater, W. & Carpenter, R.H., Snow, W., Weaver, H., White, B.T., White, J., Wyatt, T.H. and Brandon, D.
CASTLE COMBE
Gibbs, I.A., Ward & Nixon.
CASTLE EATON
Butterfield, W.
CHAPMANSLADE
Clayton and Bell, Harding, M., Horwood, Bros., Street, G.E.
CHARLTON N
Brettingham, M., Brown, L., Darley, J.
CHARLTON S
Pearson, J.L.
CHERHILL
Barry, C., Gabriel, C.H., Gabriel, S.B., Rudman, W., Rutherford, T., Slater, W., Weaver, H.
CHICKLADE
CHILMARK
Fisher, M., Harvey, F.W., Wyatt, T.H. and Brandon, D.
CHILTON FOLIAT
Blomfield, A.W., Christopher, J.T., Ferrey, B., Forsyth, W.A., Forsyth & Maule, Pilkington, W.
CHIPPENHAM
Awdry, G.C., Brakspear, H., Brakspear, O.S., Brown, R.J., Brunel, I.K., Creeke, C.C., Darley, J. & Sons, Darley, R., Downing & Rudman, Edwards, J.R., Gaye, H., Haden & Co., Hemming, A.O., Hunt, H.R., Jones, D., Llewellins & James, Matthews, H.W., Pinch, J., Pinfold, C.G., Rendell, F. and Sons, Rudman, W., Scott, G.G., Silcock & Reay, Silley, G.M., Soane, J., Stent, W.J., Thompson, J., Watts, R.P., Weaver, H., Whall, C., Willcox, W.J., Willcox, W.J., Wood, J.
CHIPPENHAM WITHOUT
Fogg, T.H., St Pancras Iron Works Co., Thompson, J.
CHIRTON

Butterfield, W., Dutch,J., Gimson, H.M., Ingleman, R., Messenger, H., Peniston, J., Seddon, J.P., Wailes, Clayton & Bell.
CHISELDON
Baverstock, W., Beswick, R.E.E., Bishop & Pritchett, Harris, J., Pinchard, B., Ponting, C.E., Reason, W., Rendell, F. and Sons, Streat, J., Webb, C.
CHITTERNE
Trotter, A.P., Wyatt, T.H.
CHOLDERTON
Bothams, A.C., Wyatt, T.H., Wyatt, T.H. and Brandon, D.
CHRISTIAN MALFORD
Provis, J.
CHUTE
Hardy & Son, Pearson, J.L.
CHUTE FOREST
Pearson, J.L., Taylor, R., White, W.
CLARENDON PARK
Dereham, E., Jones and Co., Peniston, J., Pugin, A.W.N., Wyatt, T.H.
CLYFFE PYPARD
Butterfield, W., Pace, R., Ponting, C.E., Townsend.
CODFORD
Barker, E.H.L., Eden, F.C., Imrie, G.B., Wyatt, T.H.
COLERNE
Brakspear, O.S., Goodridge, H. E., Mannings, G. & Sons, Rendell, F. and Sons, Wilson, J., Wilson & Willcox.
COLLINGBOURNE DUCIS
Blomfield, A.W., Ingelow, B., Overton, S., Street, G.E.
COLLINGBOURNE KINGSTON
Colson, J., Filer, M., Weaver, H.
COMPTON BASSETT
Hardman, J., Kennedy, G., Rendell, F. and Sons, Weaver, H., White, B.T. and Young, J., Woodyer, H.
COMPTON CHAMBERLAYNE
Soppitt, J.
COOMBE BISSETT
Fisher, F.R., Wyatt, T.H.
CORSHAM
Bellamy, T., Brakspear, H., Bromley, W. H., Butcher, R. & Son, Foster & Wood, Goodridge, H. E., Hakewill, J.H., Hansom, C.F., Hardwick, P., Kene, H., Kempe, C.E., Methuen A.P., Nash, J., Ponting, C.E., Powell and Moya, Rendell, F. and Sons, Stocking, T., Street, G.E., Weaver, H., Wyatt, J.
CORSLEY
Leachman, J., Stanley, W.H., Trotter, A.P.
COULSTON
Gane Bros., Rendell, F. and Sons.
COVINGHAM
CRICKLADE
Christian, E., Galpin, Kempe, C.E., Pace, R.
CRUDWELL
Brown, F.
DAUNTSEY
Barnes, E., Barnes,J., Bessant, J., Blanton, J., Brooks, J., Copper, D., Day, C., Greenman, G., Gunston, S., Gunston, W., Gye, Hannam, J., Hayward, J., Hind, R., Kayns, Mizen, W., Scott, W., Strong, T.,Style, A.J., Wilcox, J., Woodman, R.

DEVIZES
Abington, L.J., Anstie, J., Ash, H., Awdry, G.C., Baldwin, T., Blacking, W.H.R., Bowden, F.I., Brakspear, H., Brinkworth, R.E., Carpenter, R.C., Comper, J.N., Gamble & Whichcord, George, Gimson, H.M., Goodridge, H. E., Hansom, C.F., Hardick, E.E., Hill, W., Ingleman, R., Isborn, E.C., Kempe, C.E., Lawrence, Long, J., Maslen, L., Mullens, B., Overton, T.C., Peniston, J., Pollard, Ponting, C.E., Randell, J.A., Raymond, Rendell, F. and Sons, Roberts, E.S., Salmon, C., Scoles, A.J., Slater, W., Slater, W. & Carpenter, R.H., Troup, F.W., Ward, R.J., Weaver, H., Wilkinson, G., Wilson & Willcox, Woodyer, H., Wyatt, B.D., Wyatt, J., Wyatt, T.H., Young, J. and White, B.T.

DILTON MARSH
Aust, D., Austin, Shout & Withers, Freestone, R., Hall, R., Harding, J., Manners, G.P., Peniston, J., St Aubyn, J.P., Wilkinson and Moore, Withers, R.J., Wyatt, T.H.

DINTON
Blacking, W.H.R., Burford, J., Butterfield, W., Crook, W., Plowman, J., Wyatville, J.

DONHEAD ST ANDREW
Ponting, C.E., Tate, N.

DONHEAD ST MARY
Blacking, W.H.R., Blomfield, A.W., Butcher, R. & Son, Harding, J. & Son, Walker, W.

DOWNTON
Alexander, D.A., Blacking, W.H.R., Butterfield, W., Morris, R., Potter, R., Revett, N., Wood, J., Wyatt, T.H.

DURNFORD
Ashley, E., Blow, D.J., Bothams, Brown & Dixon, Ponting, C.E., Silley, G.M.

DURRINGTON
Hugall, J.W., Rendell, F. and Sons, Ross, W.A.

EAST KENNETT
Gane, R.

EAST KNOYLE
Aitchison, G., Blomfield, A.W., Blow, D.J., Carpenter, R.H. and Ingelow, Peniston, J., Webb, P., Wyatt, T.H. and Brandon, D.

EASTERTON
Christian, E., Sainsbury, J., Style, A.J., Weaver, H.

EASTON

EASTON GREY

EBBESBOURNE WAKE
Barnden, J., Harding, M., Singleton, G., Stent, W.J., Warren, E.P., Wilkinson, W.

EDINGTON
Blacking, W.H.R., Brakspear, H., Burgess, J., Ponting, C.E.

ENFORD
Messenger, H., Ponting, C.E., Rudman, W., Weaver, H.

ERLESTOKE
Bradie, J., Rendell, F. and Sons, Stewart, G., Street, G.E.

ETCHILHAMPTON
Weaver, H.

EVERLEIGH
Birch, J., Messenger, H., Morlidge, J., Rendell, F. and Sons.

FIGHELDEAN
Christian, E., Harding, M., Hugall, J.W.

FIRSDOWN
FITTLETON
Blacking, W.H.R., Ponting, C.E., White, W.
FONTHILL BISHOP
Harding, J., Jones, I., Wyatt, T.H.
FONTHILL GIFFORD
Adam, R., Blow, D.J., Burn, W., Devey, G., Hoare, Lane, J., Wyatt, J., Wyatt, T.H., Wyatt, T.H. and Brandon, D.
FOVANT
Clarke, W., Harding, M., Wyatt, T.H.
FROXFIELD
Christian, E., Messenger, H., Rendell, F. and Sons, Tubbs, C.B.
FYFIELD
Gabriel, C.H.
GRAFTON
Bowden, F.I., Ferrey, B., Rendell, F. and Sons.
GREAT BEDWYN
Birch, J., Boyle, R., Cundy, T., Edis, R.W., Flitcroft, H., Hardman, J., Messenger, H., Rendell, F. and Sons, Scott, G.G., Teulon, S.S., Wyatt, T.H.
GREAT CHEVERELL
Blacking, W.H.R., Woodman, W.H.
GREAT HINTON
GREAT SOMERFORD
Brakspear, H., Eden, F.C., Hakewill, J.H., Llewellins & James, Strong, W.
GREAT WISHFORD
Wyatt, T.H.
GRIMSTEAD
Pownall, F.H.
GRITTLETON
Blomfield, A.W., Clutton, H., Gibbs, A., Pinch, J., Thompson, J., Thompson, J.J., Wilmshurst, T., Wyatt, T.H.
HAM
Messenger, H., Rendell, F. and Sons, Waring & Blake.
HANKERTON
Bertodano, H.S.
HANNINGTON
Brakspear, H., Slater, W. & Carpenter, R.H.
HAYDON WICK
Drew, E.
HEDDINGTON
Brakspear, H.
HEYTESBURY
Butterfield, W., Gibbs, A., Searchfield, T.G., Trotter, A.P.
HEYWOOD
Burgess, J., Curtis, W.R.H., Eginton, H., Mundy, H.
HIGHWORTH
Alexander, G., Angell, T., Hugall, J.W., Masters, W.A.H., Pedley, W., Wilkinson, W.
HILMARTON
Butterfield, W., Salway, G.S., Street, G.E., Weaver, H.
HILPERTON
Kempe, C.E., Weaver, H., Wyatt, T.H.
HINDON
Butcher, R. & Son, Gover, W., Wyatt, T.H.

HOLT
Brown, Burgess, J., Foster, J., Ponting, C.E., Ponton, Snailum, W.W., Stent, W.J.
HORNINGSHAM
Butcher, R. & Son, Chapman, J., Smythson, R., Spicer, W., Sutton, J., Wilkinson, W., Wyatt, T.H. and Brandon, D., Wyatville, J.
HUISH
Ferrey, B.
HULLAVINGTON
Blomfield, A.W., Brakspear, H., Light, W., Ponting, C.E., Rendell, F. and Sons, Thompson, J.
IDMISTON
Pearson, J.L.
IMBER
Chapman & Sons.
INGLESHAM
Barfield, T.H., Micklethwaite, J.T., Morris, W.
KEEVIL
Adye, C.S., Ponting, C.E., Rendell, F. and Sons.
KILMINGTON
KINGSTON DEVERILL
Manners & Gill.
KINGTON LANGLEY
Barrett, J., Brakspear, H., Darley, J. & Sons, Gabriel, C.H., Goold, V., Miller, Pinnegar, J.
KINGTON ST MICHAEL
Hakewill, J.H., Pearson, J.L.
KNOOK
Butterfield, W.
LACOCK
Blomfield, A.W., Brakspear, H., Chapman, J., Gabriel, S.B., Gale, J. and Banks, G., Goodridge, H. E., Hitchcock, W., Miller, S., Peacock, K.J.R., Pritchard, J., Rudman, W., Slater, W., Wood, J., Woodyer, H., Wyatt, J.
LANDFORD
Butterfield, W.
LANGLEY BURRELL WITHOUT
Brakspear, H., Ponting, C.E.
LATTON
Angell, T., Butterfield, W., Kempe & Tower, Ponting, C.E., Roseblade, J.
LAVERSTOCK
Briggs & Gordon, Caroe, W.D., Jacob, H., Wyatt, T.H.
LEA AND CLEVERTON
Coe & Goodwin, Foster, W., Phipps, G.J., Phipps, J.C.
LEIGH, THE
Ponting, C.E.
LIDDINGTON
Cannon, J., Ponting, C.E.
LIMPLEY STOKE
Aust, D., Foster, Manners, G.P.
LITTLE BEDWYN
Gould, J., Messenger, H., Overton, S., Ponting, C.E., Randell, A.J., Rendell, F. and Sons, Weaver, H.
LITTLE CHEVERELL
Blacking, W.H.R., Cundy, T., Gimson, H.M., Martin, A.C.

LITTLE SOMERFORD
Brakspear, H., Darley, R.
LONGBRIDGE DEVERILL
Chantrey, Chapman & Sons, Eden, F.C., Gilbert, A., Harding and Elgar, Powell, R.S., Rendell, F. and Sons, Wait, Wakefield, P., Wyatt, T.H. and Brandon, D.
LUCKINGTON
Blomfield, A.W., Easton & Robertson, Kempe, C.E., Kempe & Tower, Thompson, J.
LUDGERSHALL
Anrep, B., Awdry, G.C., Blowe, H., Carrington, D., Henshaw, F., Kennedy, G., Pearson, J.L., Peniston, J., Powys, A.R., Pym, R.
LYDIARD MILLICENT
Wheeler, W.
LYDIARD TREGOZE
Lansdown & Shopland, Mantell, E.W.
LYNEHAM
Butterfield, W., Hansom, C.F., Pinnegar, C.E., Rendell, F. and Sons, Rudman, W.
MAIDEN BRADLEY WITH YARNFIELD
Chapman, J.
MALMESBURY
Awdry, G.C., Brakspear, H., Brown, F., Cockerell, C.R., Darley & Wilkinson, Darley, J., Filton, W., Goodridge, H. E., Rendell, F. and Sons, Stent, W.J., Ward, R.J.
MALMESBURY WITHOUT
Brown, F.M., Morris, W., Rosetti, D.G., Scott, G.G., Shaw, J., Weaver, H.
MANNINGFORD
Brakspear, H., Clacey, N.E., Gabriel, S.B., Newton, E., Pearson, J.L.
MARDEN
Ponting, C.E., Weaver, H.
MARKET LAVINGTON
Bowden, F.I., Christian, E., Cundy, T., Greenshields, T., Martin, A.C., Newton, E., Peniston, J., Ponting, C.E., Rendell, F. and Sons, Wyatt, T.H.
MARLBOROUGH
Aylwin & May, Baverstock,W., Blomfield, A.W., Blore, E., Blowe, H., Bodley, G.F. and Garner, Bowden, F.I., Brooke, J.W., Brunsdon, W., Cole, E., Cooper, W., Crickmay & Son, Crocker, E.H., Deane, J., Gould, J., Hammond, J., Harrison, H., Money, J., Nelson, J.M., Newton, E., Newton, W.G., Overton, S., Ponting, C.E., Rendell, F. and Sons, Roberts, D.H.P., Rogers, E., Shaw, R.N., Silcock & Reay, Stillman & Eastwick-Field, Street, G.E., Street, A.E., Ward, R.J., Webb, A., White, W., Willes, T., Wyatt, T.H.
MARSTON
MARSTON MAISEY
Brooks, J.
MELKSHAM
Adye, C.S., Brakspear, H., Curnie, W.W., Gunstone, J., Hall, L.K., Kempe, C.E., Ponting, C.E., Powell, J. & Sons, Rendell, F. and Sons, Street, G.E., Ward & Hughes, Wilson & Willcox, Wyatt, T.H.
MELKSHAM WITHOUT
Brakspear, H., Bromley, W.H., Finden, J., Money, T., Ponting, C.E., Rendell, F. and Sons, Smith, W., Wyatt, T.H.
MERE
Blacking, W.H.R., Butcher, R. & Son, Caroe, W.D., Fisher, F.R., Ponting,

C.E., Powell, J. & Sons, Scott, G.G. & Moffatt, W.B., Stent, W.J., Wadlow, H. and J., Wyatt, T.H.
MILDENHALL
Abraham, R.A., Beswick, A.E.
MILSTON
Ponting, C.E.
MILTON LILBOURNE
Lutyens, E.L., Pearson, J.L., Rendell, F. and Sons, Street, G.E., Townsend, R.
MINETY
Stillman & Eastwick-Field.
MONKTON FARLEIGH
Adye, C.S., Davis, C.E., Hicks, J., Methuen, A.P., Wantner, W.H., Wood, J., Wyatt, T.H.
NETHERAVON
Blacking, W.H.R., Briant, H. and Briant, N., Messenger, H., Ponting, C.E., Soane, J.
NETHERHAMPTON
Butterfield, W.
NETTLETON
Bromley, B., Butcher, R. & Son, Gabriel, S.B., Salway, G.S.
NEWTON TONEY
Benson, W., Sturgess, E., Wyatt, T.H. and Brandon, D.
NORTH BRADLEY
Snailum, W.W., Stanley, W.H., Wyatt, T.H.
NORTH NEWNTON
Pearson, J.L.
NORTH WRAXALL
Bromley, B., Ponting, C.E.
NORTON
Lawson, F.A., Tyrwhitt, T.
NORTON BAVANT
Edwards, E., Hardick, W.H., Harding, M., Prangley, T., Walker, W.
OAKSEY
Bridges, J.B., Lutyens, E., Rayson, T., Thomas, P.H.
ODSTOCK
Abraham, R.A., Alexander, D.A., Blacking, W.H.R., Caroe, W.D., Fowler, J., Gambier Parry, S., Harding, J., Harding, M., Jacobsen, T., Morris, R., Salvin, A., Webb, C., Wyatt, J., Wyatt, T.H.
OGBOURNE ST ANDREW
Baverstock, J., Baverstock, W.E., Brakspear, H., Butterfield, W., Ponting, C.E., Pope, J.G.
OGBOURNE ST GEORGE
Messenger, H., Wyatt, T.H.
ORCHESTON
Blake, R., Edwards, J.A., Harding, J. & Son.
PATNEY
Dyer, W., Kempe, C.E., Weaver, H.
PEWSEY
Birch, J., Brooke, J.W., Chandler, Cooper, W., Cundy. T., Gimson, H.M., Manns family, Mitchell, J., Penning, W.H., Phipps, J.C., Ponting, C.E., Street, G.E.
PITTON AND FARLEY
Aitchison, G., Blount, G.L.W., Caroe, W.D., Christian, E., Fort, A.,

Harding, M., Kempe, C.E., Wren, C.
POTTERNE
Burford, J., Christian, E., Nicholson, C., Rendell, F. and Sons.
POULSHOT
Noyes, W., Peniston, J., Ponting, C.E., Weaver, H.
PRESHUTE
Brakspear, H., White, W., Wyatt, T.H.
PURTON
Butterfield, W., Lansdown, T.S., Mantell, E.W., Phillips, J., Wilkinson, G.
QUIDHAMPTON
RAMSBURY
Caus, I., Hooke, R., Reeve, J.A., Robson, J., Sutton, B., Webb & Sutton, White, W.
REDLYNCH
Dawber, E.G., Harding and Elgar, Newman, D., Ponting, C.E., Scott, G.G.
ROUNDWAY
Adye, C.S., Chivers, W.E., Hine, G.T. & Pegg, H.C., Maslen, L., Piper, T. & W., Powell, J.G., Rendell, F. and Sons, Walker, T., Weaver, H., Wyatt, T.H.
ROWDE
Blanchard, G., Goodridge, H. E., Ponting, C.E., Rendell, F. and Sons, Weaver, H., Weaver & Adye, Wyatt, M.D.
RUSHALL
Ponting, C.E., Weaver, H.
SALISBURY
Atkinson, T., Bath, F., Beckham, H., Beckham, J., Beckham, R., Blacking, W.H.R., Blomfield, A.W., Blount, G.L.W., Bothams, A.C., Bothams, Brown & Dixon, Bowden, F.I., Brunel, I.K., Butcher, R. & Son, Butterfield, W., Cachemaille Day, N.F., Card, J., Caroe, W.D., Carter, O.B., Christian, E., Clarke, S., Clutton, H., Cockerell, S.P., Crickmay & Son, Dereham, E., Dinsley, W.H., Dyke, D.N., Elkins, E.J., Ely, N., Erlestoke, T., Farleigh, R., Fayrebowe, J., Figes, W. & Co., Fisher, F.R., Fletcher, D., Futcher, R., Gibson, J., Green, Lloyd & Son, Hall, H., Harding, J. & Son, Harding, J., Hardman, J., Hopper, T., Joseph, E.M., Kellow, J., Kempe, C.E., Lush, E., Lyons, MacFarlane & Co., Messenger, H., Mileham, G.S., Minty & Godwin, Mondey, E., Morris, W., Naish, G., O'Connor, M. and A., Oldrieve, W.T., Osmond, W., Peniston, H., Peniston, J., Philip, J.B., Pilkington, W., Ponting, C.E., Potter, R., Powell and Moya, Pugin, A.W.N., Reeve, J.A., Rendell, F. and Sons, Roe, W.H., Schafflin, R., Scott, G.G., Slater, W., Stent, W.J., Strapp, J., Street, G.E., Strong, T., Surman, R., Tarring & Wilkinson, Taylor, R., Turner, J., Webb, E.D., Wood, J., Woodyer, H., Wren, C., Wyatt, J., Wyatt, T.H., Wyatt, T.H. and Brandon, D.
SAVERNAKE
Ponting, C.E., Rendell, F. and Sons, Scott, G.G., Wyatt, T.H.
SEAGRY
Brakspear, H., Chesterton, M., Godfrey, C., Hakewill, J.H., Harrison, H.B.
SEDGEHILL AND SEMLEY
MacPhail, L., Wyatt, T.H.
SEEND
Brakspear, H., Medlicott, W.B., Style, A.J., Weaver, H.
SEMINGTON
Bailey,R., Bailey,S., Berrett, Bigwood & Co., Blacking, W.H.R., Christian, E., Clarke, J., Fryer, Gane, C. and Gane, R., Giles & Gane, Harvey, Kendall, H.E., Martin, Rawlins, Richman, G. & Co., Silverthorn, J., Snailum, W.W., Wilkes, J.

SHALBOURNE
SHERRINGTON
Prangley, J., Rogers, H.S.
SHERSTON
Christian, E., Sumsion, T.
SHREWTON
Christian, H., Fisher, F.R., Harding, J., Mount, G.E., Sleat, W., Wyatt, T.H., Wyatt, T.H. and Brandon, D.
SOPWORTH
Morris, W., Wyatt, T.H.
SOUTH MARSTON
Belcher, J., Masters, W.A.H.
SOUTH NEWTON
Lavers & Barraud, Wyatt, T.H.
SOUTH WRAXALL
Goodridge, H. E., Martin, A.C., Snailum, W.W., Weaver, H., Weaver & Adye.
SOUTHWICK
Lee, J., Millington, W., Moody, L., Ponting, C.E.
STANTON FITZWARREN
Binns, H.W., Hugall, J.W., Kempe, C.E.
STANTON ST BERNARD
Plank, J., Robbins, S., Seddon, J.P., Weaver, H.
STANTON ST QUINTIN
Brewer, Elms, W., Greenman, W., Hakewill, J.H., Knapp, J., Manning, J., Pleydell-Bouverie, B., Ponting, C.E., Rendell, F. and Sons, Smith, J., Woodman, J.
STAPLEFORD
Rendell, F. and Sons.
STAVERTON
Smith, W., Snailum, W.W., Willis, E.
STEEPLE ASHTON
Brakspear, H., Clutton, H., Giles & Robinson, Harrison, H., Lovell, T., Rawlins, W., Rendell, F. and Sons.
STEEPLE LANGFORD
Butcher, R. & Son, Carpenter, R.H. and Ingelow, Scott, G.G., Slater, W., Sleat, W., Waters, J., Wyatt, T.H.
STERT
Hakewill, J.H., Wailes, Clayton & Bell.
STOCKTON
Arnold, W., Ferrey, B., Harding, J., Lister, J., Ponting, C.E., Sarjeant, A., Wyatville, J.
STOURTON WITH GASPER
Benson, W., Buckler, J., Campbell, C., Flitcroft, H., Maggs and Hindley, Moulton & Atkinson, Odber, J., Parker, C., Webb, A., Webb, E.D.
STRATFORD TONEY
Kempe, C.E.
STRATTON ST MARGARET
Angell, T., Drew, E., Drew, W., Foden, S.P., Masters, W.A.H., Pedley and Smith, Salvin, A., Smith, T.
SUTTON BENGER
Bentley, J.F., Brick, Hakewill, J.H., Withers, R.J.
SUTTON MANDEVILLE
Alford, J., Barnes, W., Combes, C., Miles, T.B., Soppitt, J., Wyatt, T.H.
SUTTON VENY

Blount, G.L.W., Clayton and Bell, Pearson, J.L., Rogers and Booth.
SWALLOWCLIFFE
Scott, G.G. & Moffatt, W.B.
SWINDON
Architects Co-partnership, Armstrong, J., Bailey, E.N., Bertram, Bertram and Rice, Beswick, R.E.E., Beswick, R.J., Bevan, J., Binyon, B., Bishop & Pritchett, Blomfield, A.W., Bowden, F.I., Brakspear, H., Brownlow and Cheers, Brunel, I.K., Colborne, A.J., Cole, E., Cripps, W.H., Drake & Pizey, Drew, E., Drew, W., Elmbank Foundry, Ferrey, E., Gibberd, F., Gooch, D., Halliday & Roger, Harvey, E., Kempe, C.E., Lansdown, T.S., Major, G., Mantell, E.W., Masters, W.A.H., Messenger, Moore, T.L., Morgan, J.L., Morris, F., Morris, R., Nicholls, H.E., Osborne, W.R., Parker & Unwin, Phillips, J., Pike, C. & Partners, Ponting, C.E., Powell and Moya, Read, W.H., Rendell, F. and Sons, Rigby, J. and C., Roberts, E., Robertson, E., Sage, S., Scott, G.G., Scott, G.G. & Moffatt, W.B., Seddon, J.P., Shingler Risdon Associates, Silcock & Reay, Smith, H.S., Smith, J.J., Stockwell, E., Thomas, M.H., Travers, M., Turner, T., Unwin, R., Wade, F.B., Webb, E.D., Weedon, H.W. and Partners, Wilson & Willcox.
TEFFONT
Barnes, R., Butcher, R. & Son, Fowler, C., Harding, J., Moffatt, W.J., Scott, G.G. & Moffatt, W.B.
TIDCOMBE AND FOSBURY
Blomfield, A.W., Teulon, S.S.
TIDWORTH
Harding, M., Pearson, J.L., Ponting, C.E., Rendell, F. and Sons.
TILSHEAD
Prangley, T., Wyatt, T.H. and Brandon, D.
TISBURY
Beckett, A., Blacking, W.H.R., Booth & Ledeboer, Christian, E., Creeke, C.C., Hughes, R.N., Lane, J., Miles, T.B., Mowbray, Green & Hollier, Paine, J., Smythson, R., Soane, J., Soppitt, J.
TOCKENHAM
Godwin, F., Lansdown, T.S.
TOLLARD ROYAL
TROWBRIDGE
Bell, J., Blandford, H. and Smith, Bowden, F.I., Brunel, I.K., Burlison and Grylls, Collins, P.G., Davis, C.E., Davison, T., Dyer, J., Fisher, M., Gane Bros., Gane, R., Goodridge, A.S., Habershon, W.G. and Pite, Hepworth, P.D., Horwood Bros., Livesay, A.F., Long, J. & Sons, Manners, G.P., Manners & Gill, Messenger, Paull & Bonella, Powell, J. G., Powell, J. & Sons, Rendell, F. and Sons, Reynolds, E., Roberts, D.H.P., Scoles, A.J., Silcock, T.B., Smith, W., Snailum, W.W., Stanley, W.H., Stent, W.J., Talman, W., Taylor, J., Underwood, C., Warrington, Weaver & Adye, Wilkins, W.H.
UPAVON
Oliver, E.K., Rendell, F. and Sons, Seddon, J.P., Wyatt, T.H.
UPTON LOVELL
Atkinson, T., Snailum, W.W., Weaver, H.
UPTON SCUDAMORE
Street, G.E., Wyatt, T.H.
URCHFONT
Gimson, H.M., Hakewill, J.H., Messenger, H., Ponting, C.E., Randell, J.A., Rendell, F. and Sons, Roberts, E., Talman, W.
WANBOROUGH
Baverstock, W.E., Brakspear, O.S.

WARMINSTER
Blomfield, A.W., Blore, E., Butcher, R. & Son, Carson and Miller, Edwards, G., Games, T., Glascodine, J., Halliday, J.E., Hardick, T., Harris, T., Kempthorne, S., Leachman, J., Mowbray, Green & Hollier, Nicholson, C., Ponting, C.E., Ponton, A., Powell, J. & Sons, Ralph, J., Reeve, J.A., Stent, J., Stent, W.J., Street, G.E., Vialls, G., Wren, C., Wyatt, T.H., Wyatt, T.H. and Brandon, D.

WEST ASHTON
Burgess, J., Hopper, T., Lancaster, Weaver, H., Wyatt, T.H., Wyatt, T.H. and Brandon, D., Wyatville, J.

WEST DEAN
Pownall & Young.

WEST KNOYLE
Allen, J.M., Butterfield, W.

WEST LAVINGTON
Blacking, W.H.R., Kempe, C.E., Martin, A.C., Ponting, C.E., Rendell, F. and Sons, Warren, E.P., White, J.B.

WEST OVERTON
Ponting, C.E.

WEST TISBURY
Blow, D.J., Webb, E.D.

WESTBURY
Blomfield, A.W., Brakspear, H., Bruton, E., Burgess, J., Card, J., Christian, E., Eginton, H., Evans, T.L., Freestone, R., Hall, R., Harding, J., Rendell, F. and Sons, Stent, W.J., Taylor, R., White, W., Willis, F., Wyatt, T.H., Wyatt, T.H. and Brandon, D.

WESTWOOD
Brakspear, H., Jones and Atwood, Messenger, H., Peto, H., Rennie, J., Rolfe & Peter, Voisey & Wills, Wilson, J., Wilson & Fuller.

WHITEPARISH
Butterfield, W., Gimson, H.M., Harding, M., Tatham, C.H.

WILCOT
Baldwin, T., Ellis, C.W., Gimson, H.M., Money, J., Newton, W.G., Rendell, F. and Sons, Rennie, J., Style, A.J., Teulon, S.S.

WILSFORD
Weaver, H.

WILSFORD CUM LAKE
Blow, D.J., Bradell, D. and Deane, Darcy, Braddell & Deane, Webb, P., Wyatt, T.H.

WILTON
Adair, J., Blacking, W.H.R., Blore, E., Brown, B.O., Caus, I., Chambers, W., Clermont, A., Critz, E., Dawkins, T., Delamotte, V., Devall, J., Egginton, M., Elliott, T., Evans, T.L., Fisher, M., Ford, H., Gooderick, M., Gordon, G.H., Herbert, H., Humby, W., Hunt, E., James, J., Jones, D. & Jones, C., Jones, I., Kinward, T., Morris, R., Parsons, G., Philip, J.B., Pierce, E., Privett, W., Reed and Malik, Rendell, F. and Sons, Reysschoot, P.J., Ryder, R., Sabbatini, L., Smeaton, J., Townsend, R., Ward, T., Warre, E., Webb, J., Westmacott, R., Wyatt, J., Wyatt, T.H., Wyatt, T.H. and Brandon, D., Wyatville, J.

WINGFIELD
Barnden, J., Snailum, W.W.

WINSLEY
Dawber, E.G., Jones, D. & Jones, C., Pope, R.S., Preedy, F., Rennie, J., Silcock & Reay.

WINTERBOURNE
Thornton, T., Weaver, H., Wyatt, T.H.
WINTERBOURNE BASSETT
Field & Hinton.
WINTERBOURNE MONKTON
Butterfield, W., Christian, E., Gibbs, A.
WINTERBOURNE STOKE
Crook, T.
WINTERSLOW
Harding, J., Tatham, C.H., Thurston, S., Vallis & Butter, Wyatt, T.H., Wyatt, T.H. and Brandon, D.
WOODBOROUGH
Messenger, H., Wyatt, T.H., Wyatt, T.H. and Brandon, D.
WOODFORD
Blow, D.J., Darcy, Braddell & Deane, Hansom and Welch, Harding, J., Hooper & Dobbin, Ponting, C.E., Wyatt, T.H.
WOOTTON BASSETT
Barrett, T., Hardman, J., Little, R., Masters, W.A.H., Rendell, F. and Sons, Street, G.E., Wood, J.
WOOTTON RIVERS
Baverstock, W.E., Messenger, H., Mitchell, J., Street, G.E.
WORTON
Medlicott, W.B., Messenger, H., Wyatt, T.H. and Brandon, D.
WROUGHTON
Brakspear, H., Lansdown & Shopland, Masters, W.A.H., Pace, R., Rendell, F. and Sons, Wyatt, J., Wyatt, T.H.
WYLYE
Eden, F.C., Hardick, W., Hardick, W. & Son, Kempe, C.E., Peniston, J., Ralph, J., Wyatt, T.H. and Brandon, D.
YATTON KEYNELL
Street, G.E.
ZEALS
Scott, G.G.

FOLEY, SON, & MUNDY,

H. MUNDY, F.S.I. VALUER, appointed by the Wiltshire County Council, under the Finance Act, 1894.

H. G. FOLEY.

AUCTIONEERS, BREWERS', INNKEEPERS' AND

General Valuers, Land Surveyors,

Land and Estate Agents, Accountants, &c.

ESTABLISHED 1845.

THE MART, MANVERS ST., TROWBRIDGE.

THE SALE OF ALL KINDS OF

FREEHOLD & PERSONAL PROPERTY, Etc.,

Effected either by Public Auction or Private Contract.

CATTLE & AGRICULTURAL STOCK of all descriptions Sold at their Cattle-ring in Trowbridge Market or elsewhere.

Valuations of Estates, Land, Houses and Furniture; Brewers', Innkeepers', and other Business Stocks;
Made either for Sale, Transfer, Administration, or Mortgage.

Estates, Houses, and all kinds of Buildings, etc., accurately Surveyed and planned. Plans of Property put on Deeds. Estates wound up. Debts collected. Business in any of the above branches undertaken in any part of the Kingdom.

FOLEY, SON & MUNDY beg to notify that they have a large and commodious Sale Room in MANVERS STREET, TROWBRIDGE, and are prepared to receive Furniture and Effects for Periodical Sales. They also beg to inform their Agricultural friends that they have a permanent CATTLE RING, COVERED SHED, and PENS in the TROWBRIDGE MARKET, where they hold Sales on the fixed Market days (every alternate Tuesday). Entries are solicited for these Sales, and those received on the Thursday previous to the Market, will have the advantage of advertisements in the local papers FREE OF COST. The best attention given and SETTLEMENT IMMEDIATELY AFTER SALE.

Agents to the Alliance Assurance, Fire, Life, and Hail, The Eagle Life, The London Guarantee and Accident, and the Accident Insurance Companies; and the Employers' Liability Assurance Corporation.

4, Cricklade St., King William St. & Prospect Pl.,
OLD SWINDON.

ESTABLISHED 1856.

G. R. HENLEY,

Plumber, Decorator,

BUILDER,

CARPENTER & GENERAL CONTRACTOR.

GRAINING, WRITING & MARBLING.

Beer Engines, Water Closets, Deep Well Pumps, Spirit Racks, &c.,
ERECTED AND REPAIRED ON THE MOST APPROVED PRINCIPLE.

HOUSE DECORATING IN THE FIRST STYLE.

PAPER HANGINGS.

A great variety kept in Stock.

Estimates given for General Repairs.

Experienced Workmen sent to any part of the Country.

ADVERTISEMENT.

Mʀ. W. H. READ,

Architect and Surveyor,

SWINDON.

SURVEYOR TO SWINDON PERMANENT BUILDING SOCIETY.

OFFICES:

CORN EXCHANGE.

SWINDON
Marble, Granite,
AND
STONE WORKS,
Bath Road, SWINDON,
AND AT
CLIFTON STREET,
NEW SWINDON,
(Close to the Cemetery Entrance).

GEORGE WILTSHIRE,
BUILDER,
General Contractor, Lime Burner, &c.
BUILDING IN ALL ITS BRANCHES.
ESTIMATES SUPPLIED.
Oak and Dry Laths, Dry Board and Plank always on hand.
Timber cut by Steam Power to any Scantling required.

DRAINAGE PIPES & MATERIALS KEPT IN STOCK.
ROMAN AND PORTLAND CEMENT.

Bath Stone, Fire Goods, Stone Steps, Cattle Troughs, Slates, Tiles, &c.,
Wholesale and Retail.

LIME OF THE BEST QUALITY.

At Reasonable Prices, at the Quarry Works, or can be delivered to Order.
if required.